Tom Mawhinney's argument is one that every strategy leader needs to hear. A Distinct Advantage reframes competitive advantage as a leadership discipline, not a set of results. It effectively demonstrates how leadership, strategy and technology is the modern equation for creating and sustaining advantage in the new era. The book is both practical and challenging. It reinforces, in a structured way, the critical need of discipline as a leader and in how we need to apply rigour in how we do things. The message is urgent and deeply relevant. It is the work of staying competitive today.

- PATRICK KRAMER, CEO of BDO Global

A Distinct Advantage *is a grounded, timely book for leaders operating in real complexity. Tom Mawhinney makes a compelling case that modern competitive advantage lives in the alignment of leadership, strategy, and technology, not in isolated initiatives or short-term wins. What I appreciate most is that this is not theory for theory's sake. It gives executives a clear way to diagnose friction, accelerate decision-making, and build momentum that actually lasts.*

- CAREY JAMES, Founder & Executive Director, Brand Alchemy

Finally, a book that treats technology not as a back-office function but as an equal partner in competitive advantage. A Distinct Advantage *doesn't just diagnose what's broken in traditional strategy—it offers a coherent, actionable framework for what comes next, an equal partnership between leadership, strategy and technology. This is Mawhinney's wake-up call for boards still treating digital transformation as someone else's job.*

- CARRIE PURCELL, AI Advisor, Speaker, Podcast Host,
Author, Top 50 Women in Business

A Distinct Advantage *is an essential read for any leader seeking to understand what truly drives sustained performance in a modern organization. Tom Mawhinney makes a convincing case that competitive advantage no longer stems from strengths of the past, but from the dynamic interplay of contemporary leadership, differentiated strategy, and integrated technology. The framework is clear, actionable, and purpose built for leaders trying to look at challenges through an updated lens. At a time when every organization must rethink how it competes, this book offers one of the most practical and forward-looking views available.*

JIM KRAHN, Growth, Operations & Emerging Technology Expert

A DISTINCT ADVANTAGE

Defining Modern Competitive Advantage
for Your Environment

A DISTINCT ADVANTAGE

Defining Modern Competitive Advantage
for Your Environment

TOM MAWHINNEY

THIN LEAF PRESS

Publication Data
Names: Mawhinney, Tom, Author
Titles: *A Distinct Advantage*

ISBN: 978-1-968318-35-2 (paperback) | 978-1-968318-34-5 (eBook)

Business, Leadership, Strategy, Technology
Editors: Dhanliza Cellona
Thin Leaf Press
Los Angeles Book Cover

Design by 100Covers Interior
Formatting by Dindo B. Sanguenza

THIN
LEAF

TABLE OF CONTENTS

INTRODUCTION

The quarterly business review began like hundreds before it. Slides were crisp, forecasts current, and the conversation confident. Revenue was slightly ahead of plan, margins a little thinner, and a few renewals were taking longer than expected, but no one viewed the results as cause for concern. The team adjourned with the familiar assurance of leaders who believed they were managing well: steady performance in an unpredictable market.

They were managing well, yet managing well no longer meant competing well.

The leadership team still prided itself on discipline. Meetings were structured, information thorough, decisions reasoned. But the tempo was slow. Issues were analyzed, re-analyzed, and deferred until consensus formed. The pace that once signaled rigor now produced hesitation. By the time the organization aligned, the market had already shifted again.

The strategy was equally deliberate. It promised reliable growth through proven channels and reinforced the company's reputation for consistency. Yet the market was rewarding something different. New entrants were redefining value around expertise and speed, not size or history. The plan still looked solid on paper, but the edges that once differentiated it had dulled.

Technology sat quietly beneath the conversation, still treated as support rather than as a source of advantage. Systems were stable, but they reflected how the business used to operate. Each digital initiative added cost and complexity at the margins rather than momentum at the core. The tools were modern; the mindset around them was not.

Individually, none of these conditions seemed urgent. Together, they explained why a respected company now felt ordinary. The organization wasn't broken; it was built for a world that no longer exists. Its leadership cadence, strategic logic, and technology posture all came from an era when advantage could be built gradually and defended for years. *That era is over.*

Scenes like this unfold every quarter in capable organizations. These are not companies in crisis; they are the ones that appear stable until stability becomes their greatest liability. The instincts that built their strength, discipline, reliability, and patience remain admirable but have become insufficient in an environment defined by speed, transparency, and continual reinvention. The world has changed, and the playbook that once created advantage now quietly erodes it.

This book begins with that recognition. It explores what now drives competitive advantage and why the familiar instincts of leadership, strategy, and technology no longer hold their edge. It introduces how these three variables must work together if a company hopes not merely to keep pace but to define the pace of its industry.

When Advantage Collapses

The meeting you just read about could have taken place in almost any organization. Across industries, the pattern is familiar: The strengths that once defined stability now quietly undermine it. What leaders were trained to treat as advantage, including brand reputation, operational scale, distribution reach, and accumulated experience, no longer hold the ground they used to. The very tools that once created momentum have become the weight that slows it.

For much of the last century, those tools worked because the environment allowed them to. Markets rewarded repetition more than reinvention. Companies built patiently, optimized relentlessly, and defended what they owned. Advantage was something you could engineer through assets and efficiency, then guard with discipline. Time and clarity were the hidden variables of success.

Those conditions have disappeared. Today, change arrives from every direction at once. Technology accelerates the flow of information and collapses distance. Workforces and customers expect meaning and immediacy, not just quality. Investors, regulators, and communities judge impact as closely as performance. These forces overlap and amplify one another until the environment becomes not just faster but more complex. It is an interconnected web of pressures that refuses to slow down. The systems that once favored size and steadiness now reward organizations that are able to interpret and act through complexity in real time.

In this world, advantage no longer comes from possession; it comes from progression. Scale, distribution, and brand still matter, but their value depends on the capabilities that shape them. Leadership defines pace and credibility. Strategy determines focus and integrity. Technology connects and accelerates them both. Together, they form the infrastructure through which every other strength gains relevance.

Throughout this book, modern competitive advantage refers to these connected disciplines of leadership, strategy, and technology, each redefined for the realities of today's environment. When practiced together, they keep organizations aligned as the world accelerates. This is the process of modern advantage. It turns awareness into alignment and alignment into motion, providing leaders with a way to refine and sustain performance across cycles of change.

That is the structural shift at the heart of this book. Advantage has not disappeared; it has changed form, moving from what organizations own to how they operate. The question for leaders is no longer how to protect what they have, but how to keep the processes that create advantage alive. Understanding that shift is where modern competitive advantage begins, and where, ultimately, a distinct advantage takes shape.

The Modern Equation

If the familiar sources of strength no longer create advantage, the next question is obvious: What does? Across industries, organizations that continue to move forward share a common pattern. Their success is not the result of a single breakthrough or a fleeting idea but of mastery across three enduring disciplines: how they lead, how they decide, and how they connect capability to action. *These disciplines, leadership, strategy, and technology, form the variables of a modern equation for advantage.*

This equation does not describe a fixed formula; it defines the variables within a dynamic process: the ongoing work of creating, connecting, and renewing advantage. Leadership, strategy, and technology have always existed, but the way they generate performance has changed. Each now operates at a different pace, under greater transparency, and in constant interaction with the others. Individually, they build strength; together, they create momentum.

Leadership remains the first test of an organization's relevance, but what it demands has shifted. Modern leaders must build the capabilities to interpret complexity, align direction, and mobilize people and systems at the pace of change. Credibility depends less on position and more on fluency in the forces reshaping industries: digital acceleration, stakeholder expectations, and societal transparency. The role of leadership is to turn uncertainty into focus and ensure that the organization moves in time with change, not behind it.

Strategy is still the blueprint of performance, but its center of gravity has moved. Where it once depended on scale and efficiency, strategy today depends on clarity and distinction. It is not about predicting the future but about designing for it, making deliberate choices that can flex as conditions evolve. A modern strategy is alive: It connects aspiration with adaptability and treats renewal as discipline, not disruption.

Technology has become the connective force that enables both leadership and strategy to work at their intended pace. Emerging capabilities—including artificial intelligence, automation, advanced analytics, and digital platforms—now define how value is created, delivered, and measured. Technology no longer sits beside the business; it is the environment in which the business operates. It amplifies clarity when leadership and strategy are aligned and exposes weakness when they are not. It sets the tempo for the entire enterprise.

Each discipline matters on its own, but their collective power lies in how they interact. Leadership provides direction and credibility, strategy provides clarity and distinction, and technology provides speed and reach. *When these forces move together, the pursuit of advantage becomes a process that is dynamic, repeatable, and resilient.* Modern competitive advantage begins with excellence within each discipline and endures through their alignment. The chapters ahead explore these disciplines in depth and show how, when modernized and connected, they become the engine of a distinct advantage.

When Alignment Becomes the Edge

Recognizing what drives modern advantage is only the beginning. The real work lies in elevating each discipline: how organizations lead, how they make and defend choices, and how they use technology to meet the realities of today's marketplace. Alignment among them matters, but only after each has evolved. When leadership still relies on hierarchy, when strategy still prizes efficiency over distinction, or when technology still operates as a back-office utility, synchronization changes nothing. It can only amplify what already exists. *Excellence within the variables is the foundation, and synchronization across them is what turns that excellence into sustained momentum.*

Modern organizations operate in environments where every decision touches every other. Leadership sets direction, but without a clear strategy, the direction fragments. Strategy provides focus, but without technological capability, it cannot move fast enough to matter. Technology accelerates possibility, but without leadership and strategy, it magnifies confusion as easily as progress. Alignment is the discipline that keeps these forces in step, the difference between motion and traction, between activity and advantage.

The benefits of alignment appear subtle at first and decisive over time. Integration shortens reaction time. Shared context reduces rework. Clarity compounds. When the three disciplines move together, the organization learns faster, executes cleaner, and renews itself more easily. The work becomes lighter not because there is less to do, but because energy is no longer lost to friction between parts of the system.

True synchronization, however, is never a finish line. It is a discipline to be maintained under pressure. Markets shift, strategies evolve, and technology advances faster than governance can write policy. The process of advantage must therefore stay in motion, modernizing each discipline while keeping the whole synchronized. Misalignment among modern systems still erodes performance, but alignment among legacy ones only accelerates decline.

That is why alignment is the edge. It multiplies what modernization makes possible. Organizations that master both—excellence within and synchronization across—create a form of advantage that is dynamic, repeatable, and distinct. The next section explores the cost of failing to keep that process in motion and why, in a world defined by pace and complexity, standing still is the most expensive decision of all.

The Cost of Standing Still

Recognizing what must change is not the same as changing it. Many organizations stop at that edge. They see the forces reshaping their markets—faster competitors, new technologies, rising stakeholder expectations—and still hesitate. The hesitation feels rational: wait for clarity, wait for proof, wait for the environment to settle before committing to a new path. Yet in the modern landscape, waiting is rarely neutral. It is, in most cases, the quietest form of decline.

Advantage erodes differently now. It does not vanish in a single disruption; it decays in the intervals between recognition and response. Each pause in the process allows momentum to leak away. Every quarter that a strategy remains unrefreshed, a competitor finds a new way to differentiate. Every year that technology decisions are deferred, the organization's infrastructure hardens around outdated assumptions. Every time leadership delays renewal, culture absorbs the hesitation. Standing still no longer preserves stability; it freezes the very process that creates advantage.

For many companies, the cost of delay hides behind numbers that still look acceptable. Performance appears steady until the process stops producing learning, and by then, the gap is structural. That is the paradox of the modern landscape: by the time evidence of decline is visible, recovery requires reinvention. Organizations that stay ahead act before certainty arrives. They shorten recognition time, reward experimentation, and treat progress as movement rather than milestones. Progress begins with motion, not perfection.

The purpose of modernization is not to chase every new trend but to build the capacity to keep the process of advantage in motion. Modern advantage belongs to leaders who can adapt without starting over, who treat adjustment as discipline rather than reaction. They understand that the real risk lies not in change itself but in resisting the pace at which change occurs.

The next section explores how this capacity for deliberate motion turns a modern foundation into something rarer still:

7

a distinct advantage. It looks beyond the universal equation of leadership, strategy, and technology to how that process takes shape within each organization's unique environment.

From Formula to Distinction

Recognizing the equation that defines modern advantage is essential, but understanding it as part of a living process is what makes it useful. The equation provides clarity, and the process gives it life. Every organization must translate that process into its own environment: its markets, its culture, its risk posture, and its ambitions. The logic is universal, but the expression never is. *Advantage becomes distinct only when the disciplines of leadership, strategy, and technology take shape within the realities of a specific context.*

No two organizations start from the same place. Some possess strong leadership credibility but struggle to renew strategy. Others have bold strategies that falter because technology and execution lag behind. Many have invested heavily in new tools yet continue to apply them through outdated ways of working. Each situation demands a different entry point. The challenge is not to copy what others have done but to adapt the same process to the conditions that define your own world.

That is why the pursuit of advantage is both structured and personal. The equation travels because the process behind it is universal; it provides a repeatable way to understand where advantage now comes from and how it is sustained. But its application requires judgment. Industry dynamics, customer expectations, regulatory realities, and cultural norms all shape how the three disciplines should take form. The objective is not conformity to a model but coherence within your environment.

Leaders who grasp this distinction treat the equation as guidance, not prescription. They start where their leverage is highest, evolve the remaining disciplines in motion, and build alignment as they go. Some begin by redefining leadership expectations, others by sharpening strategy or embedding technology directly into the work. What matters is the trajectory, progressing from isolated excellence toward a connected process that holds together under pressure.

This final step in the introduction points to where the book goes next. The chapters ahead define the equation in detail, show how to modernize each discipline, and explain why alignment multiplies their effect. The final part turns to application: how to translate the foundation into a distinct advantage within your environment and how to make the process yours.

How This Book Works

This book follows a sequence that mirrors the real process of building and sustaining advantage. It begins with awareness, moves through modernization, and ends with application. Each part brings together several chapters that explore a common theme while contributing to a single progression: understanding what has changed, redefining what advantage now requires, and learning how to apply it within your environment.

Part I – The Changing Foundation of Advantage

These opening chapters show why the traditional playbook no longer works and examine how organizations behave under the new conditions: leaders who adapt, followers who imitate, and those who stall. Together, they build the case for redefining advantage as a dynamic process.

Part II – The Equation for Modern Competitive Advantage

This part defines the three variables in depth and shows how their modernization forms the new foundation of advantage. Individual chapters focus on each: how leadership must evolve, how strategy must differentiate, and how technology must enable speed and connection, culminating in how they operate together as one continuous process.

Part III – Excellence Within and Synchronization Across the Variables

This part shows how advantage is sustained once the foundation is built. It explains why both excellence within and synchronization across leadership, strategy, and technology are required to keep performance renewable, and what happens when either discipline fades or the rhythm between them breaks.

Part IV – When Modern Competitive Advantage Becomes a Distinct Advantage

The final set of chapters turns the universal process into practice. They illustrate how to apply the approach within your organization's realities—including market position, culture, and risk posture—and how to sustain modernization as an ongoing discipline rather than a one-time transformation.

The book is meant to be read sequentially but not rigidly. Readers can enter wherever the most leverage exists. Some will begin with leadership, others with strategy or technology. What matters is movement across the three. Every part—and every chapter within it—builds toward the same goal: a repeatable, modern process for creating and sustaining advantage in your world.

The Work Ahead

The work of building advantage has never been more demanding— or more achievable. Every organization, regardless of its history, faces the same reality: Advantage can no longer be inherited or assumed. It must be created, refined, and sustained as a process in motion. Some will use this book to strengthen what once set them apart; others will use it to build advantage for the first time. The starting points differ, but the discipline is the same.

The path forward is not about reacting faster. It is about acting with greater clarity. The ideas that follow are designed to help you see the forces shaping your environment and to equip you with a process for turning that awareness into progress. Each part builds on the last, showing how modern leadership, strategy, and technology form the foundation of advantage, and how—when modernized and aligned—they lift performance to a higher level of relevance and resilience.

This book will challenge assumptions, confirm instincts, and—if it does its job—change the questions you ask about your own organization. It does not offer a checklist or a single method—because advantage cannot be templated. Instead, it provides an approach, a way to think, to sequence, and to sustain the work of creating something distinct in your world.

The invitation is simple. Begin where you have influence. Modernize what must evolve. Connect what must work together. Move deliberately from recognition to action, and keep the process in motion. Advantage is no longer a position to hold; it is a process to master. The foundation is modern, but the advantage ahead will be distinct because it will be yours.

PART I

The Changing Foundations of Advantage

CHAPTER 1:
A DIFFERENT GAME
ENTIRELY

The Erosion of Traditional Advantage

For decades, the sources of competitive strength were largely mechanical. Scale created efficiency, distribution guaranteed reach, and brand secured loyalty. These levers rewarded those who could build patiently and defend relentlessly. Once established, they offered stability; the game was to extend advantage, not reinvent it. Leaders were trained to manage resources, not to rethink the rules.

Those rules no longer apply. The traditional levers of advantage have not simply weakened; they have shifted position in the competitive landscape. Scale now slows adaptation, distribution adds fixed cost where flexibility is needed, and brand amplifies every inconsistency between promise and delivery. The same assets that once protected incumbents now expose them. What used to accumulate value through repetition now loses value through inertia.

The reasons for this shift lie deeper than market turbulence or faster competitors. The environment itself has changed shape. Technology has erased distance and compressed time. Customers operate across boundaries that companies still treat as separate industries. Information moves instantly, expectations update continuously, and entire categories converge in real time. The pace is visible, but pace is only the surface. Beneath it sits a new structure of competition, one that rewards clarity, integration, and traction more than it rewards possession of assets.

In this new structure, the traditional tactics of advantage depend on broader, systemic forces. Scale, distribution, and brand no longer create strength on their own; they reflect the strength of the forces that shape them: how an organization leads, how it makes and defends choices, and how it applies technology to connect intent to execution. Advantage has not vanished; it has simply moved from the assets companies control to the capabilities that make those assets work together. *Leadership, strategy, and technology have become the true levers of modern advantage because they determine how every tactical element performs.* When those forces evolve, the entire process advances. When they stagnate, even the largest balance sheets erode quietly underneath.

Evidence of this inversion is visible across every sector. In financial services, digital challengers win not through lower costs but through transparency and immediacy. In manufacturing, automation and analytics transform efficiency programs into ongoing cycles of learning. In healthcare, AI-driven diagnostics and virtual access redefine how patients choose providers. In retail, loyalty built over decades dissolves the moment a new entrant delivers greater convenience or personalization. These examples differ in form but share the same cause: Organizations built for stability now compete in environments defined by motion.

This erosion is not a temporary correction; it is a structural change in the logic of competition. Markets will not slow, and customers will not revert to earlier expectations. The challenge for leaders is no longer to perfect the familiar but to recognize that the foundations of advantage have moved. Understanding that movement—and the capabilities behind it—is where modern competitive advantage begins.

The next section examines one of the most common misreadings of this shift: the belief that the problem is speed. Speed is visible and measurable, but it is not the core issue. The deeper problem is complexity, the number of forces now moving at once, and the ability of leadership, strategy, and technology to operate together within it.

Why Speed Alone Doesn't Explain It

Executives often describe the modern marketplace as if the clock simply ticks faster: Customers expect more immediately, competitors launch sooner, and technologies mature at record pace. The observation is accurate but incomplete. Speed is visible, measurable, and easy to discuss in meetings, yet it is also the least interesting part of what has changed. The deeper shift is structural: It is the number of forces now moving at once and the degree to which they amplify one another. Complexity, not velocity, is what now overwhelms traditional models of advantage.

In earlier decades, change tended to arrive in sequence. An economic shock would be followed by recovery; a technology would diffuse slowly across sectors; new competitors would emerge within recognizable boundaries. Today, those forces arrive together, interact instantly, and compound in unpredictable ways. A shift in regulation triggers investor pressure. A technology breakthrough alters customer expectations before the product even ships. Social narratives move faster than companies can issue statements. Organizations are no longer managing discrete challenges; they are navigating overlapping networks of consequence.

This concurrency has rewritten the basis of competition. In energy, renewable innovation, geopolitical realignment, and new financing models have collided to redefine what cost leadership even means. In logistics, automation, sustainability mandates, and shifting trade flows have turned what used to be linear supply chains into adaptive ecosystems. In entertainment, audience behavior, platform algorithms, and creator economics now evolve together, forcing companies to rethink both production and ownership models. These examples differ in context but reveal the same reality: Advantage can no longer be planned sequentially because the environment no longer moves sequentially.

The old playbook treated the world as a set of controllable levers: Optimize one, then the next. The modern environment behaves more like an interdependent process: Touch one element and the entire system shifts. Leadership instincts built for hierarchy struggle in networks. Strategies built for control falter in uncertainty. Technologies designed for efficiency lack the adaptability that this environment demands. *Organizations that focus solely on running faster within old structures simply accelerate their misalignment.*

Understanding this distinction matters because it reframes what progress looks like. The challenge is not to outrun disruption but to operate coherently amid constant interaction. The organizations that manage this coherence are not necessarily faster; they are clearer about how information moves through them, how decisions connect, and how quickly learning turns into action. What separates them from the rest is not a stopwatch; it is the process underneath their pace.

The compression of markets and the simultaneity of change share a single catalyst: technology. It is technology that has collapsed distance, multiplied interdependencies, and made pace permanent. Yet it is only part of the picture, an accelerator of other forces such as stakeholder expectations, workforce mobility, and social transparency. To understand why the modern environment behaves this way, and why advantage now shifts so quickly, we have to understand how these forces interact through technology. The next section explores that interaction and the influence it has had on the balance of advantage.

Technology Has Reshaped the Balance of Advantage

The past few years have reordered the elements of competition. Technology has become indispensable but not independent. *It no longer sits beneath leadership and strategy as support; it now stands beside them as an equal force.* Every advantage that organizations once built through scale, distribution, or brand now depends on how effectively these three forces, direction, capability, and design, work together.

This reordering is recent and profound. Digital infrastructure, artificial intelligence, automation, and data connectivity have turned pace into permanence. Information moves instantly, expectations update continuously, and industries converge in real time. Yet technology is only part of the story. It connects and amplifies the others. Shifting workforce expectations, expanding stakeholder scrutiny, and new societal responsibilities have all added layers of complexity that technology binds together. The environment has become not merely faster but fully linked.

For leaders, this has created a new balance of power. Strategy can no longer claim primacy because its execution speed is constrained by technological readiness. Technology cannot lead on its own because capability without direction creates noise. Leadership sits between them, translating purpose into motion. The relative weight of each force has changed, but their dependence on one another has never been greater.

This is why technology deserves particular attention before we discuss the broader foundation of modern advantage. It is the most visibly transformed of the three disciplines and the one that now dictates the rhythm of change. But its influence is catalytic, not sovereign. It magnifies both strength and weakness. It makes good leadership visible and poor strategy obvious. Its rise does not diminish the importance of modernization elsewhere; it makes that modernization urgent.

Understanding this rebalanced landscape is essential before we examine how advantage is now built. The next section illustrates these forces in practice through a single company, showing how leadership, strategy, and technology—distinct in purpose yet inseparable in effect—behave inside a real organization.

Introducing Meridiem

Before exploring how these forces interact inside real organizations, this book follows one company in particular.

Meridiem, a medium-sized global provider of industrial technology and services, serves as the reference point throughout the chapters that follow. Its story is dynamic rather than static, appearing at different stages as the company confronts the same pressures explored throughout this book.

The purpose of using a single company is not to offer an ideal model but a working mirror, a way to see how leadership, strategy, and technology behave in motion. Through Meridiem, you will see how advantage erodes, rebuilds, and eventually becomes habit.

Meridiem's First Signal

The quarterly review at Meridiem opened with the same numbers that had defined its success for years. Revenue held steady. Margins met expectations. The board's questions were familiar, and the answers came easily. Across continents, the company's products and maintenance services powered factories, distribution hubs, and public infrastructure. By every traditional measure, Meridiem was performing as it always had. Yet behind the routine, something felt slower, heavier. The pace of the business had begun to change.

Customer decisions were coming faster, but internal ones were not. Product teams delivered on schedule, but the commercial organization struggled to respond to shifting demand patterns. Field-service data showed contracts taking longer to renew, and projects that once closed easily now required new layers of justification. The leadership team noticed but explained it away: Competitors always moved quickly until the economics caught up with them. The room's comfort came not from denial but from habit.

Meridiem's scale had long been its strength: distribution reach, brand reputation, and long-term customer relationships built over decades. Yet these advantages were becoming harder to manage. Each new product introduction demanded more coordination across geographies, more sign-offs, more compromise. The leadership team still prided itself on discipline, but what once felt like precision now felt like drag.

Strategy meetings, once animated by growth ambitions, had begun to sound defensive. The company's plans remained consistent: Defend share in core markets, expand cautiously where customer density justified it, and focus investment on incremental efficiency. The strategy read well on paper but felt increasingly detached from the pace of the market. Smaller entrants were capturing attention by integrating software and analytics into traditional equipment, something Meridiem had discussed for years but never executed at scale.

Technology investment was substantial but fragmented. Each division had modernized its own systems, often to good effect, yet no one could describe how those investments connected. The CIO's dashboard showed progress; the business units saw complexity. Technology was still viewed as infrastructure rather than as the environment in which the business operated.

What Meridiem called discipline was beginning to look like delay. Decisions waited for consensus, data waited for validation, and opportunities waited for annual budget cycles that no longer matched the market's tempo. The leadership team could sense the friction but struggled to name its cause. They were working as hard as ever, perhaps harder, yet progress felt slower.

The first signal came quietly. During a planning session, one executive summarized the company's challenge in a single sentence: "We're running last year's playbook in a season that doesn't exist anymore." The room fell silent. It was not a crisis; revenues were stable, customers loyal, but something had shifted.

Meridiem's leaders began to see that their advantage was not failing; it was fading. The playbook that had built the company's strength was still producing results, but smaller ones each quarter. The problem was not execution; it was the approach. Meridiem was managing the business as if advantage were a position to defend rather than a process to renew.

That realization, subtle and unfinished, was Meridiem's first signal. It marked the point where awareness began but motion had not yet followed, a moment of recognition that would eventually lead the company to question how advantage must now be built, not held.

The New Stakes of Advantage

Meridiem's story is not unique; it is the quiet pattern of modern decline. Advantage no longer fades at the edges of markets; it thins from within organizations that wait too long to renew themselves. The risks rarely appear as sudden collapses. They show up as slower decisions, cautious plans, and confidence built on numbers that no longer reflect reality. Performance still looks acceptable until the process that creates it stops producing learning.

The new stakes of advantage are measured in time and confidence. Every delay between recognition and action gives complexity a head start. Each year that leadership, strategy, and technology evolve separately, the gap between potential and performance widens. The issue is rarely awareness; it is alignment and motion. Companies do not lose advantage because they fail to see change coming; they lose it because they pause when the process demands movement.

Modernization has become the minimum requirement for relevance. It is no longer a choice to update how an organization leads, decides, and connects capability; it is a condition for staying in the game.

From Conditions to Response

The conditions that now define competition confront every organization, but not all respond in the same way. Some respond with urgency, translating awareness into motion. Others imitate, borrowing visible moves without changing the architecture underneath. A few remain still, convinced that the environment will stabilize or that their legacy will protect them. The difference among them is not resources or intent; it is how they interpret change and how quickly they convert recognition into renewal.

Every organization recognizes signals of change, but not every organization reacts the same way. The instinctive response is often to accelerate activity: launch new initiatives, refresh messaging, invest in technology. Yet speed without clarity rarely improves competitiveness. Movement is useful only when it aligns the disciplines that create advantage: leadership, strategy, and technology. When these evolve in isolation, the result is motion without progress.

The real test is translation: how effectively an organization converts external pressure into internal momentum. Leaders who manage that translation treat awareness as a starting point, not an endpoint. They shorten the distance between recognition and decision, between decision and action. They build a cadence that keeps the process of advantage in motion.

For others, the pattern looks different. They see the same conditions but interpret them through the lens of the past. They tighten controls, add layers of review, or wait for clearer evidence before moving. In doing so, they preserve activity but lose adaptability. Their response to complexity is management, not modernization.

These patterns separate organizations into three familiar postures. There are the leaders who move first and learn fastest, shaping advantage as they go. There are the followers who copy visible moves but struggle to replicate the coherence behind them.

And there are the stalled, confident in their current playbook, waiting for stability that will not return.

The next chapter explores these postures in more depth. It reveals why leadership, strategy, and technology, when treated as connected processes, determine which posture an organization ultimately occupies and how quickly it can change position.

Chapter Summary

Advantage has long served as a measure of sustained performance, but its nature has fundamentally changed. What once lasted for decades now dissolves in months. The foundations executives were taught to rely on—scale, brand, distribution, and relationships—are no longer fortresses. The modern environment is defined by forces that collide rather than arrive in sequence, leaving advantage temporary by default.

This chapter introduced both the modern foundation of advantage and the company that will illustrate it throughout the book. Meridiem enters here as a practical reference point, showing how traditional strength begins to erode when the environment outpaces the organization's pace. Its early signals provide context for how leadership, strategy, and technology behave once the forces of modern competition take hold.

In today's marketplace, any advantage that renews itself stands out more sharply than ever, and when it continues to renew, it becomes unmistakable. The central question, then, is how organizations respond inside these conditions. That divergence, between those who adapt, those who follow, and those who stall, is the focus of *Chapter 2: The Leaders, the Followers, and the Stalled.*

CHAPTER 2:
THE LEADERS, THE FOLLOWERS, AND THE STALLED

Diverging Paths Under Pressure

The forces described in Chapter 1 are not selective. Every organization feels them. Technology accelerates change and resets expectations. Geopolitical shifts ripple through supply chains. Capital markets compress planning horizons. Stakeholders demand more visibility, faster. Employees raise the bar on culture and meaning.

These conditions are universal, but outcomes are not. Some organizations move ahead decisively, converting disruption into a springboard for advantage. Others react, adjusting at the margins but never quite breaking through. Still, others stall, busy but misdirected, clinging to playbooks that no longer apply.

If the conditions are shared, why are the outcomes so divergent? The difference lies not in the environment but in the responses organizations adopt under pressure. Across industries, three consistent patterns emerge: the leaders, the followers, and the stalled. The leaders are fewer, and their pattern is clearer: excellence within each of the three disciplines and synchronization across them.

By contrast, there are many more ways to fall short. Followers and stalled organizations show up in multiple guises, which is why they receive more attention here. Most organizations will find themselves in one of these two camps, and that recognition is the first step toward moving forward.

- *The leaders modernize leadership, strategy, and technology, and keep them synchronized in motion,* converting pressure into leverage that becomes modern competitive advantage.

- *The followers advance in one or two disciplines but not all three,* leaving them short of the goal line.

- *The stalled misdiagnose the game altogether,* waiting for clarity that never comes, or confusing motion with momentum.

These are not permanent identities. Organizations can move between them depending on how they respond. But in the modern environment, where cycles compress and disruption collides, where you land matters because it determines whether modern competitive advantage is created and renewed, pursued but missed, or quietly eroded.

The Leaders: Converting Pressure Into Modern Advantage

The leaders are not immune to disruption. They face the same volatility, the same collapsing cycles, and the same expectations that challenge everyone else. What separates them is how they convert those forces into leverage.

For these organizations, clarity at the top is not optional. Leadership alignment establishes a shared understanding of where the market is heading and why. Strategy channels that understanding into a differentiated position competitors cannot easily match.

Technology embeds and accelerates it, not as a project or department but as the connective process of the enterprise. The three disciplines reinforce one another, creating a cadence of learning and momentum.

Consider a healthcare system in the American Midwest. Facing rising costs, workforce shortages, and virtual-first entrants, many regional providers defaulted to incremental efficiency programs. This system's leaders chose differently. They agreed on a single ambition: to be the most accessible provider in the region, physical or digital. Strategy shifted from maximizing facility utilization to re-imagining patient access. Technology followed with a telemedicine platform integrated with AI-driven triage and predictive scheduling. Within two years, the organization had not only stabilized but also expanded its referral base. Competitors were left reacting to its playbook.

A second example shows the same pattern in professional services. As AI-enabled competitors offered faster, cheaper insights, a global consultancy reframed itself around human-machine integration. Its consultants would not compete with algorithms but through them, applying judgment enhanced by data. Leadership alignment was explicit. Strategy shifted toward advisory services defined by influence and decision quality. Technology was embedded into every engagement, improving delivery instead of being sold as an add-on. While peers discounted rates, the firm expanded margin.

The leaders move with speed not because they rush but because they are coherent. Leadership clarity prevents delay. Differentiated strategy makes trade-offs explicit. Technology converts direction into action. They rarely wait for perfect conditions. They act, adjust in motion, and in doing so, they create the modern advantage others must respond to. Their advantage is not the product of size or luck; it is the outcome of disciplined modernization within each discipline and continuous alignment across them.

The Followers: Progress Without Integration

The followers are not weak organizations. Many are well managed, disciplined, and filled with capable people. They see the same changes in their environment and often respond quickly. But their progress is partial. They move forward in one or two disciplines without achieving synchronization across all three.

Consider a technology firm that built an AI-driven analytics platform years ahead of its competitors. The breakthrough was real, but leadership was divided on where to focus: large enterprises or mid-market clients. Strategy fractured, messaging grew inconsistent, and sales defaulted to familiar accounts. The company's technological advantage never translated into market traction. This is the classic follower pattern: progress in one discipline undermined by drift in the others.

A European bank offers another example. Its leadership team rallied around a clear strategy of personalized client advice. Culture and credibility were strong. But its outdated technology—with slow interfaces and fragmented transactions—clashed with rising expectations for seamless digital experiences. Competitors delivered instant mobile-first interactions while the bank's promise collapsed under the weight of its technical inertia. Leadership and strategy were sound; technology—the necessary accelerator—was missing.

A third case from retail shows how leadership alignment can falter even when investment is bold. A multinational grocer poured capital into e-commerce but could not agree on whether physical stores or digital channels should lead. The resulting half-measure strategy was neither bold enough online nor renewed in-store. Resources spread thin, results diluted, and advantage eroded. The grocer's issue was not foresight but fragmentation, a lack of integration across disciplines that turned progress into static.

Followers can remain competitive for a time. They often appear successful, sustaining respectable performance year after year. That is what makes this position deceptively safe. But the gaps compound. Customers notice. Competitors exploit the seams. The organization's position is revealed not by its effort but by its outcomes. And over time, followers remain just short of breakthrough, close enough to see it, never integrated enough to achieve it.

The Stalled: Misdiagnosing the Game

The stalled are the most vulnerable not because they lack capability but because they misread the environment. They often appear busy, with full calendars, new projects, and ambitious slogans, but the activity is misdirected. Initiatives multiply, dashboards fill with metrics, and leaders point to motion as evidence of progress. Yet without integration, motion becomes noise. In today's market, hesitation is no longer neutral; it accelerates erosion.

A global consumer-goods company once dominated premium shelf space through brand loyalty and distribution reach. When digital-native competitors began capturing share online, the company responded with a burst of isolated initiatives: a loyalty app here, an influencer campaign there, a pilot e-commerce store that never scaled. Each effort looked impressive in isolation, but none was anchored to a coherent strategy or connected by technology. Internally, teams worked hard; externally, the brand looked active. Sales flattened, and within three years, the once-dominant label had become a commodity. This is one way organizations stall, mistaking tactical activity for strategic progress.

A second case comes from the insurance industry. A regional carrier long known for service quality assumed its reputation would insulate it from digital disruption. When new entrants began offering instant, app-based policies, leadership delayed modernization, arguing that customers valued personal relationships over convenience. By the time they launched a digital platform, competitors had already locked in market share. The carrier's service ethic remained admirable, but its relevance had evaporated. This is how delay becomes decision, waiting for clarity that never arrives.

Manufacturing tells a different version of the same story. A supply-chain equipment company invested heavily in automation, convinced that efficiency alone would sustain its edge. Leadership celebrated productivity gains while overlooking that competitors were connecting automation to customer data and predictive analytics. The company's factories became faster but not smarter. It excelled at output, not insight, and lost contracts to rivals that could anticipate problems before they occurred. Here, legacy strength, operational excellence, had quietly turned into liability.

The pattern is consistent. The stalled are not inert, nor are they incapable, but they misdiagnose the game they are in. Some scatter their energy across disconnected efforts. Others delay, hoping disruption will stabilize. Still, others double down on legacy strengths long after their value has peaked. In all cases, they drift further behind with each cycle, not for lack of effort but for lack of synchronization. They confuse discipline with rigidity and management with momentum. And in a world defined by compression and convergence, those are costly mistakes.

Meridiem: Stalled in Plain Sight

When we last saw Meridiem, the company had begun to sense that its familiar playbook no longer matched the pace of the market. Yet rather than question the way they were working, its leaders looked for proof that the business was still modern enough to compete.

The leadership team gathered for what was billed as a progress review. On the screen was their much-heralded client platform, a polished dashboard promoted as evidence of digital relevance. It promised real-time project tracking and greater transparency. The CEO opened with confidence, pointing to encouraging feedback from several early clients.

Beneath the optimism, unease simmered. Sales leaders admitted they struggled to explain the platform's real value. Clients noticed the dashboard but cared more about missed deadlines and inconsistent service. Employees viewed it as extra work rather than an improvement to theirs. Competitors, meanwhile, were embedding technology into the core of their delivery models, not displaying it at the margins.

As the meeting continued, fractures widened. One group pushed to double down, convinced scale would validate the investment. Another argued for cost-cutting to preserve margin. A third leaned on legacy strengths. "Clients don't care about digital bells and whistles," one executive insisted. "They care about reliability." Each argument sounded rational on its own, but together, they revealed a deeper issue. Meridiem's leaders no longer shared a common view of what advantage meant.

The board was unconvinced. Directors pressed for clarity. Was Meridiem chasing growth or defending its base? How did the platform connect to strategy? Leadership responded with polished presentations and confident language, but the contradictions were impossible to miss. One director summarized the mood privately: "They look busy, but they're not moving forward."

Inside the company, middle managers felt the confusion most. Some were told to promote the platform aggressively; others were told to reassure clients that nothing was changing. The mixed messages fed quiet cynicism. One manager joked that progress reviews were all slides and no substance. Morale dipped as employees sensed the drift.

Externally, stakeholders began to question whether Meridiem's steadiness had turned into inertia. Clients who once prized reliability now wondered if the company could still innovate. "We don't need another dashboard," one customer said. "We need faster delivery." The market was moving, but Meridiem was not.

Meridiem was not leading. It was not even following. It was stalled in plain sight, vulnerable not because it lacked talent or ideas, but because its leaders could not align on how to respond. The appearance of progress had replaced the reality of it. What had begun as a quiet signal of drift was now a visible stall, the point where comfort had overtaken motion.

Why Modern Advantage Is So Rare

Meridiem's experience is not unusual; it reflects a pattern repeated across industries. Organizations face the same conditions but achieve very different outcomes. The difference lies not in the environment but in the ability to modernize and align the disciplines that create advantage. Few manage to do both at the same time, and even fewer sustain it.

Across industries, the pattern is unmistakable.

- The leaders modernize and align leadership, strategy, and technology, converting disruption into advantage.
- The followers make progress in one or two disciplines but fall short of coherence across all three.
- The stalled misdiagnose the game, mistaking activity for progress and delay for discipline.

The deeper reality is this: *Modern competitive advantage is rare, not because organizations lack resources or intelligence, but because coherence across these disciplines is difficult to achieve and harder still to sustain.*

In earlier decades, advantage was rare because it was slow to build. Scale, distribution, and brand demanded years of investment. Once secured, they endured. Today, the opposite is true. Advantage is rare because it erodes quickly. Conditions collide, cycles compress, and competitors replicate features overnight. A position of strength that once lasted decades can vanish in quarters.

Why is alignment so elusive? Because leadership, strategy, and technology each move on different clocks. Leadership cycles revolve around credibility, culture, and people. Strategy cycles follow markets, positioning, and economics. Technology cycles are faster still: New tools appear before leaders can evaluate them, before strategies can adapt, and before organizations can absorb the change. Each discipline speaks a different language, moves at a different pace, and is driven by different incentives. Without deliberate alignment, divergence becomes the default.

Even capable organizations slip into follower or stalled patterns. They pursue excellence in one discipline while neglecting the others. They reward visible activity instead of meaningful progress. They delay decisions in search of perfect clarity, not realizing that waiting is itself a decision, one that accelerates erosion.

This rarity also sharpens the value of modern advantage. Because so few organizations can maintain both excellence within and alignment across the three disciplines, those that do stand out immediately. Customers feel it in better experiences. Competitors see it in lost deals. Markets reward it in reputation and valuation. When advantage functions as a continuous process rather than a static position, it not only lasts longer; it compounds.

Leaders should pause here. Ask candidly: Which of these patterns best reflects us today? Are our leadership, strategy, and technology reinforcing one another in motion or moving at different speeds, pulling the organization apart? The discipline of answering honestly is the first step toward building modern advantage.

The Pursuit of Modern Advantage

Chapter 1 revealed the collapse of familiar conditions, the erosion of traditional levers that once defined strength. Chapter 2 exposed how organizations respond to that pressure: the leaders who create genuine advantage, the followers who progress without integration, and the stalled who drift behind. Together, they show that recognition alone is not enough. In the modern environment, advantage depends on the ability to translate recognition into renewal and renewal into rhythm.

The lesson is simple but demanding. Advantage in today's market cannot be built through motion alone. It requires deliberate modernization within leadership, strategy, and technology, and continuous alignment across them. The leaders who pull ahead do not succeed because they are larger, wealthier, or luckier. They succeed because they treat these disciplines as interconnected processes, kept in motion at the pace of disruption. Excellence within each creates strength, and alignment among them converts that strength into momentum.

For the followers and the stalled, the gap is rarely intelligence or intent; it is their approach. Without coherence across disciplines, even capable organizations lose traction. With it, advantage becomes dynamic, repeatable, and resilient, a process that renews itself as conditions change.

This is the pursuit of modern advantage: the ongoing work of modernization and alignment in rhythm with a world that refuses to stand still. The next chapters define this foundation in detail, examining how leadership, strategy, and technology each evolve and how, together, they create a modern advantage that endures.

Chapter Summary

Having seen how organizations diverge under the same pressure, one truth stands out: Advantage today is not determined by conditions but by response. The same volatility that erodes some companies propels others forward. The difference lies in how leadership, strategy, and technology evolve and align under stress, whether they move as isolated functions or as connected disciplines within a single process of renewal.

The patterns are consistent. The leaders modernize each discipline and synchronize them in motion, converting disruption into leverage. The followers advance partially, progressing in one or two areas but failing to achieve coherence across all three. The stalled misdiagnose the game, mistaking activity for progress and delay for discipline. These are not fixed identities; organizations move between them depending on how quickly they translate recognition into modernization and modernization into alignment.

This is the essence of modern competitive advantage. It is not a state to defend but a process to master—dynamic, repeatable, and strengthened through coherence across leadership, strategy, and technology. Few organizations sustain it for long, which is why it remains rare and valuable. The next section, *Part II: The Equation for Modern Competitive Advantage,* opens with *Chapter 3: The Core Equation of Modern Competitive Advantage,* which defines this foundation in detail and shows how leadership, strategy, and technology each create strength on their own, and how—when integrated as a single process—they become the engine of advantage that endures through constant change.

PART II

The Equation for Modern Competitive Advantage

CHAPTER 3:
THE CORE EQUATION OF MODERN COMPETITIVE ADVANTAGE

Beyond the Noise

The faster the environment moves, the easier it becomes to mistake activity for progress. Dashboards update in real time, campaigns launch in rapid succession, and new initiatives appear designed to prove momentum. Yet beneath this noise of constant motion, something quieter is happening: advantage begins to dissolve. Data now moves faster than decisions. Metrics multiply while meaning thins. In a world defined by speed and complexity, even capable organizations lose traction because they are managing the surface instead of the engine.

The result is familiar. Culture campaigns expand while culture itself weakens. Customer-experience programs multiply while customers drift. Brand sentiment spikes and collapses in the same quarter. Each effort measures coherence, but none creates it. The visible indicators of advantage, such as employee engagement, customer loyalty, and brand admiration, are outcomes, not causes. Managing them directly is leading from lagging indicators. When leaders focus on what is easiest to measure, they lose sight of what is hardest to sustain.

Speed and complexity distort focus. When markets shift weekly and data refreshes hourly, visibility feels like control. Executives chase what can be monitored rather than what must be mastered. The dashboard becomes a substitute for direction. As the organization learns to read numbers instead of patterns, performance becomes a series of reactions rather than a process of renewal. Activity accelerates while true progress stalls.

The pattern is exhausting. Teams respond faster but learn less. Meetings fill with information that proves awareness but rarely creates movement. Organizations chase employee-survey surges or customer-satisfaction gains, convinced that motion at the edges signals strength at the core. Yet these metrics track coherence; they do not build it. A logistics company can celebrate record delivery times and still erode profitability if leadership, strategy, and technology are out of step. A university can post record enrollment and still lose relevance if its academic model, digital systems, and governance operate on different clocks.

What sustains advantage today lies deeper. It is the ability to generate progress continuously rather than perform it temporarily. That ability depends on three key variables that operate together: leadership, strategy, and technology. Leadership creates clarity and adaptability when conditions collide. Strategy designs distinction and coherence that customers can feel. Technology provides the connection and pace that allow both to function at speed. Each must perform at a high level on its own, but their true power emerges only when they move in concert.

This alignment is the new foundation of performance. It is what converts motion into momentum and renewal into rhythm. Modern competitive advantage is no longer defined only by excellence within single disciplines but by the interaction among them. Understanding that interaction is the purpose of what follows: the core equation of modern competitive advantage.

Defining the Equation of Modern Advantage

Modern organizations operate in an environment defined by speed, transparency, and continual change. The traditional foundations of strength, such as control, scale, and efficiency, no longer provide stability. They create motion but not necessarily progress. Modern advantage depends instead on three variables that now determine how performance is built and sustained: leadership, strategy, and technology. These variables have always defined how organizations perform, but the modern environment has redefined how they must evolve and function together. Each has evolved from a fixed construct into a living capability that operates continuously and in relation to the others. Together, they form the structure of modern competitive advantage.

The first variable, *contemporary leadership*, defines how organizations create direction and credibility in motion. It begins with awareness that the environment has changed. Modern leaders face forces of speed, transparency, and interdependence that demand new responses. They build from skills to broader competencies, combining adaptability, judgment, and learning to stay credible under pressure. Leadership now depends on fluency more than authority and connects people, purpose, and pace so the organization moves as one.

The second variable, *differentiated strategy*, defines how organizations create coherence and distinction. It determines where to focus and how to win. Modern strategy functions as a living framework that is clear in intent, flexible in form, and reinforced through consistent decision-making. It ensures that every action contributes to a single pattern of value creation that customers and competitors can both recognize.

The third variable, *integrated technology*, defines how organizations create connection and speed. It links decision to execution, enabling precision, insight, and learning across the enterprise. Technology no longer serves as background infrastructure; it has become the environment in which leadership and strategy operate at the required pace. When aligned with intent and design, it accelerates progress. When managed as an isolated pursuit, it fragments it.

These three variables form the core of modern competitive advantage. The next three chapters examine each in greater depth, explaining how contemporary leadership acts as the catalyst, how differentiated strategy provides the blueprint, and how integrated technology serves as the enabler that keeps the enterprise in motion. Together, they form the foundation of modern advantage, but their full power depends on how they operate in concert.

Each variable must achieve depth and credibility on its own before interaction among them can create lasting value. Leadership must provide direction and confidence. Strategy must provide focus and coherence. Technology must provide connection and speed. This is the foundation of excellence within, the mastery that builds credibility and depth inside each variable. From that foundation, synchronization across the variables creates rhythm, the alignment that turns individual strength into shared momentum. Modern advantage depends equally on both dimensions. Depth builds credibility; rhythm keeps it alive.

When excellence within and synchronization across coexist, the organization begins to perform differently. Leadership provides clarity. Strategy turns that clarity into coordinated direction. Technology enables decisions to move through the enterprise at the speed of relevance. Work feels connected. Meetings become shorter because priorities are clear before they begin. Adjustments happen in real time because information and judgment move together. Progress compounds through alignment rather than intensity.

This dual discipline of excellence within and synchronization across keeps advantage renewable. It allows leadership, strategy, and technology to reinforce one another rather than compete for control. When the variables remain in rhythm, change no longer feels like disruption; it becomes movement the organization can interpret and use.

The equation redefines what strength means in a modern enterprise. Advantage now depends less on assets and more on alignment, less on ownership and more on rhythm. The stronger each variable becomes, the more they amplify one another. Contemporary leadership keeps adaptation possible. Differentiated strategy keeps coherence visible. Integrated technology keeps pace sustainable. Together, they transform motion into momentum and complexity into flow.

Modern advantage depends on both depth and rhythm, but few organizations maintain both for long. The next section examines why this balance is so difficult to preserve and how it determines whether advantage endures or erodes.

Excellence within each variable is necessary but insufficient

For much of the last century, mastery within a single capability could sustain advantage for years. Scale, efficiency, and reputation created margins wide enough to protect companies from the consequences of drift. A strong leader could rely on experience to correct mistakes. Even a relatively forward-thinking strategy could deliver returns while execution caught up. A technological breakthrough could define a market long before competitors responded. That world has disappeared. The forces that once moved in sequence now collide at speed, exposing the limits of strength achieved in isolation.

Excellence within each variable remains essential. It builds credibility and depth inside every organization and forms the foundation of performance. Leadership, strategy, and technology must each demonstrate competence on their own before they can reinforce one another. Yet in the modern environment, that foundation no longer guarantees progress. The challenge is not whether organizations can achieve mastery, but ultimately whether they can connect it. The real test of performance lies in how well leadership, strategy, and technology move together in real time.

Isolated excellence often hides behind strong results. A well-crafted strategy can temporarily offset slow technology adoption. A visionary leader can create energy that disguises execution gaps. A sophisticated platform can generate data that looks like progress, while the organization loses coherence beneath the surface. Each can deliver success for a time, but without synchronization across the variables, the gains remain temporary. Over time, energy disperses and the enterprise begins to work harder but not faster, with each improvement in one area revealing weaknesses in another.

Consider a global consumer-goods company that built its reputation on operational efficiency. Its strategy was rigorous and its cost discipline legendary. Leadership cultivated a culture of precision, and technology kept production seamless. Yet as customer expectations shifted toward personalization and digital engagement, the organization's strengths became boundaries. Strategy remained sound, but it could not keep pace with an environment that demanded coherence among leadership, strategy, and technology. The company's precision turned into rigidity, and its consistency became friction. By the time it responded, smaller but more synchronized competitors had already redefined the market.

LEGO offers the opposite story. After years of overextension in the early 2000s, the company rediscovered strength through deliberate alignment of the same variables. Leadership clarified a singular purpose: to inspire and develop the builders of tomorrow. Strategy refocused on innovation through play and community. Technology became the bridge, linking physical creativity with digital design and analytics that connected fans, designers, and operations in real time. Each variable reached a world-class standard, but it was their synchronization that created renewal. LEGO transformed from a manufacturing company into a global creative ecosystem. Its success was not the product of size or invention but of coherence that turned strength into motion.

The difference between these stories is not ambition but rhythm. In markets where decisions collide rather than queue, isolated excellence decays faster than weakness in motion. Leadership without strategic clarity generates movement without direction. Strategy without credible leadership stalls in execution. Technology without either amplifies confusion instead of traction. Speed and complexity compress the distance between these variables so tightly that weakness in one instantly exposes the others. Depth remains the foundation of advantage, but it must now operate as part of a broader system that keeps leadership, strategy, and technology connected.

This compression explains why many organizations struggle to recognize their own drift. Performance indicators often remain positive even as coherence erodes. Strategy documents grow more detailed, technology investments expand, and leadership communication increases. Activity becomes a substitute for rhythm. The organization looks busy because it is compensating for misalignment. Progress requires more effort because decisions no longer reinforce one another. Energy leaks through the seams of disconnected work.

The discipline of modern advantage reverses that pattern. Excellence within each variable builds depth and credibility, but synchronization across them converts that depth into momentum. When leadership, strategy, and technology evolve together, they reinforce one another continuously. The organization learns faster than it operates. Strategy stays relevant because leadership interprets change quickly. Technology enables those adjustments without losing continuity. The enterprise becomes self-correcting, capable of staying coherent even as its environment shifts.

Inside organizations that achieve this balance, advantage feels different. Meetings end sooner because alignment precedes conversation. Investment decisions are simpler because priorities are shared. Employees understand how their work contributes to both purpose and performance. Technology becomes the means of learning, not the record of it. Coherence generates confidence, and confidence accelerates progress. The organization stops reacting to change and begins shaping it.

Make no mistake, the strength of modern competitive advantage lies in mastering two connected disciplines: depth within and rhythm across. Depth provides mastery; rhythm ensures relevance. Without mastery, rhythm has nothing to coordinate. Without rhythm, mastery loses momentum. Together, they create resilience, the capacity to adapt continuously without losing coherence. The organizations that understand this do not chase transformation; they sustain it. They treat rhythm as a capability in its own right, as important as any product, process, or market position.

The next section explores why this balance has become so rare and why maintaining it requires more than good management. It demands leaders who can interpret convergence, organizations that can operate at multiple speeds, and systems designed to stay coherent under continuous pressure.

The Rarity of Modern Advantage

Coherence has become the scarcest condition in modern performance. It exists when excellence within leadership, strategy, and technology connects across them, creating rhythm that sustains clarity and momentum. Leadership, strategy, and technology are visible everywhere, yet sustained alignment among them has become exceptional. The variables themselves have not weakened; what has changed is the environment that surrounds them. Pace, transparency, and interdependence have removed the space organizations once relied on to recover from misalignment. In a world that moves as a single network, even capable institutions lose rhythm faster than they can restore it.

This scarcity is not the result of declining talent but of rising complexity. Forces that once moved sequentially now collide at speed. Economic cycles, stakeholder expectations, social narratives, and technological advances interact continuously, amplifying one another's effects. Partial alignment can no longer hold. Each variable must now evolve in rhythm with the others if the enterprise is to remain coherent.

Acceleration exposes that fragility. Technology increases the pace of every decision, expanding both reach and consequence. Leadership choices cascade instantly through systems and markets. Strategic clarity must travel at the same speed to keep direction intact. When any one of the three lags, coherence thins and performance begins to depend on effort rather than rhythm. Speed magnifies potential and vulnerability in equal measure.

The compression of time has changed what strength looks like. In the traditional landscape, a dominant strategist, a visionary leader, or a breakthrough technology could sustain advantage for years. Today, excellence within remains essential, but specialization without synchronization creates friction. Advantage no longer resides solely inside functions; it depends on both mastery within them and the coherence that connects them. Those connections are fragile because they are human, built through trust, communication, and shared context that must be renewed continuously.

This is why coherence has become the true measure of ambition. It reflects not only depth within disciplines but humility across them, the willingness of leadership to listen, of strategy to adapt, and of technology to serve purpose rather than pace. Excellence within remains the foundation, yet synchronization across now determines endurance. The discipline required to hold the two together has outgrown the comfort of structure and the tempo of management. It demands constant interpretation and deliberate rhythm.

Organizations that master this discipline operate differently. They do not chase transformation; they sustain it. Alignment is not an event but a behavior, reinforced daily through leadership clarity, strategic focus, and technological maturity. Their reward is stability that moves with its environment, a form of resilience that grows stronger under pressure.

The next section returns to Meridiem, where the absence of this rhythm becomes visible. Its experience shows how capable organizations lose momentum not through failure but through drift, and how recognizing that drift becomes the first signal of renewal.

Meridiem: When the Equation Breaks

Meridiem's efforts to prove modernization had revealed something deeper: Activity was rising, but alignment was breaking down.

The leadership team now knew something was wrong. Client retention was slipping, growth had slowed, and morale had become uneven. Their instinctive response was to reach for visible fixes. A culture campaign launched, engagement scores rose briefly, and a brand refresh followed with new messaging and a confident tone. The numbers looked stable, but momentum continued to fade. The organization was strengthening signals of progress while the connection between them weakened.

The board pressed for clarity. "Where is the real problem?" one director asked. The answer arrived quietly. Three of the company's core functions—leadership, strategy, and technology—were no longer moving at the same pace. Leadership still promised reliability. Strategy continued to defend the familiar. Technology ran its own modernization effort, producing updates that impressed internally but failed to translate into client impact. Each was competent on its own, yet none were connected. Progress had become friction, and the harder the company worked, the heavier the motion felt.

Every initiative that followed repeated the same pattern. Culture programs focused on sentiment rather than coherence. Technology projects delivered new capabilities but little integration. Leadership reviews emphasized effort instead of traction. Meridiem was not failing through negligence or lack of intelligence; it was failing through disconnection. The variables that once complemented one another had become sources of tension. Activity multiplied while rhythm thinned.

The moment of recognition came during a quarterly review. The CEO described the company's condition as steady but slower. The chair responded by drawing a simple line on the whiteboard, labeling one side internal motion and the other shared rhythm. The team listed their current initiatives under motion: engagement, satisfaction, productivity, client feedback. They hesitated when asked to identify what held those efforts together. What began as a management discussion turned into a moment of truth. The organization had been managing activity rather than coherence.

That realization shifted the conversation. Meridiem's leaders saw that their performance issues were not about capability but about connection. Leadership, strategy, and technology were strong in isolation but misaligned in motion. The cadence of the enterprise had slowed because its variables no longer moved together. The problem was rhythm, not resources.

In the weeks that followed, the leadership team began to question how renewal could start. They recognized that modernization within their environment could not begin with technology or strategy; it had to start with leadership. Only leadership could restore rhythm and re-establish the conditions for coherence. That recognition marked a turning point. Meridiem's advantage had not vanished; it had drifted. The first step toward recovery would not be reinvention but reconnection.

Chapter Summary

Chapter 3 introduced the core equation of modern competitive advantage. Strength today depends on both excellence within and synchronization across the variables that create it. The three critical variables—contemporary leadership, differentiated strategy, and integrated technology—must each achieve mastery within their own domains while staying synchronized in motion. Excellence within builds credibility; synchronization across converts that credibility into momentum. Together, they create coherence, the quality that turns activity into advantage and progress into renewal.

Excellence remains essential, but it is no longer sufficient. In a world where forces move simultaneously, isolated mastery quickly erodes. Leadership without strategic clarity loses direction. Strategy without credible leadership stalls in execution. Technology without either produces speed without traction. Modern advantage emerges only when these disciplines reinforce one another continuously, creating the rhythm that keeps organizations coherent as conditions change.

The next chapter, *Contemporary Leadership: The First Variable of Modern Advantage,* explores how leaders evolve to meet modern forces, developing the skills that combine into competencies and keep organizations coherent in motion.

CHAPTER 4 – CONTEMPORARY LEADERSHIP: THE FIRST VARIABLE OF MODERN ADVANTAGE

The Catalyst of Modern Advantage

Leadership is the most visible of the three variables of modern advantage but also the most misunderstood. Entire industries exist to measure it, teach it, and critique it. Yet despite this constant scrutiny, many organizations still cling to outdated assumptions about what leadership is and how it works. They rely on models rooted in stability and hierarchy, as if credibility and tenure alone could carry an organization through markets defined by speed and collision.

That world has passed. In the modern environment, leadership must evolve in step with the forces shaping competition. It must move at the pace of technology cycles, absorb the pressures of geopolitical shifts and stakeholder demands, and create clarity amid volatility. Leaders who fail to evolve with these conditions do not simply fall behind; they become liabilities. Their organizations slow, drift, and eventually cede ground to competitors that are better aligned with the realities of the market.

This is why leadership stands as the first variable of modern advantage. It activates the entire equation. Without leadership, strategy remains theory and technology becomes noise or novelty. Leadership supplies the credibility that aligns teams, the clarity that focuses resources, and the adaptability that keeps strategy and technology moving together. It is the spark that determines whether the other variables can gain traction at all.

Modern leadership begins with the disciplined development of essential skills, the capabilities that allow leaders to interpret complexity, communicate with precision, and act decisively under pressure. Some of these skills will be new; others will need to be redefined for a different pace and scale of competition. They form the building blocks of the competencies that produce clarity, adaptability, and credibility. Yet skills alone do not create momentum. Advantage emerges only when those skills combine into broader competencies that hold under pressure, allowing leadership to translate direction into action and to keep pace with the environment it serves.

Modern leadership is therefore not about charisma, position, or tenure. It is about the deliberate evolution of skills into competencies that can guide strategy and technology in rhythm with one another. It is about leading in context, adapting without losing direction, and stewarding the enterprise through continual change. That is the work of contemporary leadership, and it is where we begin.

Leadership Must Evolve in Step with the World

Leadership has always mattered, but in the modern environment, its role has fundamentally changed. Earlier eras rewarded stability. Leaders were expected to provide consistency, reinforce established practices, and steward resources through long growth cycles. Tenure and experience once signaled reliability, and credibility was built through continuity. Those qualities still matter, but they are no longer enough. The conditions leaders face today move too quickly and interact too widely for yesterday's playbook to hold.

Markets now move at a pace that outstrips traditional approaches to leadership. Technology compresses cycles and resets expectations. Stakeholders, from investors to employees, demand more transparency and faster responses. Customers shift loyalties quickly, drawn to organizations that deliver clarity, connection, and purpose as naturally as they deliver products. In this environment, leadership that does not evolve becomes an obstacle rather than an asset. It slows the organization at the very moment when coherence and pace matter most.

The contrast between legacy leadership and modern leadership is stark. Legacy leadership treats credibility as something earned once and protected through control. Modern leadership recognizes that credibility must be re-earned continuously through judgment, communication, and adaptability in motion. It begins with the disciplined development of skills that enable leaders to understand their environment and guide their organizations within it. These skills, ranging from decision-making and translation to problem-solving and communication, form the groundwork of competence. But they create impact only when they evolve together, combining into capabilities that allow leaders to interpret complexity, maintain alignment, and sustain progress under pressure.

Modern leadership, therefore, depends on more than experience. It demands a willingness to expand skill sets, to relearn how to lead in environments defined by technology, interdependence, and transparency. The ability to synthesize complexity, to communicate with precision, and to act decisively under uncertainty is no longer a differentiator; it is the minimum requirement for relevance. Leadership that resists this evolution not only loses pace but also risks pulling strategy and technology out of rhythm, weakening the very system it is meant to guide.

Consider the case of a global consumer brand once led by a widely respected executive known for operational discipline. For years, that discipline kept margins strong and competitors at bay. But as digital-native challengers reshaped the customer experience, the same disciplined leader resisted change, treating innovation as a distraction. What had once been a strength became a liability. The leader had not expanded the skill set or adapted existing ones to fit a new context, and the organization paid the price.

By contrast, other leaders have embraced this evolution. When Salesforce grew from a niche software provider to a global platform, Marc Benioff reframed the company's leadership ethos almost immediately. He recognized that credibility was not about personal reputation but about the organization's ability to stay relevant in a changing world. He emphasized adaptability, stakeholder engagement, and the cultivation of new capabilities across his team. That shift not only differentiated Salesforce strategically but reinforced its leadership credibility with customers, employees, and investors.

Modern leadership is a continual act of renewal. It asks leaders to combine timeless instincts with new capabilities, to revisit what effectiveness means, and to guide organizations in rhythm with the pace of change. Leaders who fail to evolve risk slowing the system around them. Those who do evolve become the catalysts who turn volatility into clarity and motion into advantage.

Modern Skills Create Modern Competencies

Much of the conversation about leadership still centers on individual skills such as communication, decision-making, empathy, and resilience. These remain essential, but they represent only part of what modern leadership demands. The challenge for today's leaders is not simply acquiring skills but combining them in ways that create coherence and momentum under pressure.

Effective leadership begins with a foundation of essential skills across professional, personal, and technology domains. These include visioning, translation, decision-making, problem-solving, communication, motivation, coaching, authenticity, emotional intelligence, empathy, curiosity, adaptability, resilience, and the technology skills of awareness, objectivity, and application. Together, these capabilities provide the foundation for clarity, understanding, and sound judgment across the organization.

Possessing the skills is only the beginning. What differentiates contemporary leaders is how they integrate them into competencies that sustain rhythm when conditions collide. Communication becomes credibility when strengthened by authenticity. Decision-making becomes adaptability when joined with problem-solving and curiosity. Empathy becomes cultural alignment when combined with motivation and emotional intelligence. Resilience becomes trust when reinforced by consistency and authenticity. Competencies emerge when these skills work together, translating intent into shared direction and collective motion.

During the pandemic, many leaders demonstrated empathy by acknowledging employee stress and uncertainty. Only a few translated that empathy into alignment, combining communication, authenticity, and problem-solving to sustain productivity and trust. Both groups possessed the skill; however, only one built the competency to turn empathy into traction.

Modern leadership, therefore, requires deliberate synthesis. The combinations that matter most depend on the forces at play, including technology cycles, stakeholder scrutiny, market convergence, regulatory change, and cultural transformation. Common competency baselines include the ability to synthesize complexity, engage stakeholders in real time, and align culture with strategy. Each relies on a blend of underlying skills working in concert rather than in isolation.

Competence also has context. A startup founder, a hospital administrator, and a regional bank CEO will rely on different combinations, yet all must evolve beyond technical mastery into integrated leadership capability. The specific mix changes, but the principle endures: Modern advantage demands leaders who can combine skills into competencies that create clarity, adaptability, and motion across the enterprise.

Modern leadership development must therefore move beyond skill accumulation. It must challenge leaders to test and integrate what they know, to expose gaps, and to build competencies that connect leadership, strategy, and technology into a single rhythm of performance. Skills create potential; competencies convert that potential into momentum.

Competence in Context

Competence is always contextual. The skills and combinations that define effective leadership vary depending on the environment and the pressures shaping it. A startup founder navigating rapid growth faces a different challenge than a hospital administrator coordinating patient care or a regional bank executive balancing risk and regulation. Their conditions differ, but the expectation is the same: to adapt faster than the environment shifts and to sustain coherence when familiar patterns no longer hold.

Modern advantage does not reward static expertise. It rewards leaders who can extend, refine, and recombine their skills as conditions evolve. Decision-making, communication, and problem-solving remain essential, but how those skills are expressed, and in what combination, depends entirely on context. The most capable leaders are not those with the longest list of abilities but those who can draw from the right ones at the right moment, aligning professional, personal, and technological capabilities in motion.

In healthcare, for example, competence demands a balance between adaptability, authenticity, and technological fluency. Leaders must translate complexity into clarity while maintaining empathy for patients, clinicians, and communities. They make decisions under pressure and across disciplines, guided by a mix of evidence, ethics, and humanity. In technology sectors, curiosity and resilience define competence, supported by sound judgment and data literacy. Leaders in these settings must embrace innovation while resisting the seduction of speed for its own sake. In financial services, translation, emotional intelligence, and problem-solving often determine success. Effective leaders interpret complexity for others, converting regulation, analytics, and risk into clear direction that sustains confidence under scrutiny.

Each industry reveals different combinations, but the same principle applies. Leadership competence is never static and never isolated. Context determines emphasis; integration determines effectiveness. The leader of a logistics firm undergoing automation, for instance, must merge clarity with adaptability, ensuring that digital tools enhance rather than replace judgment. A manufacturing executive leading a sustainability transition may lean on communication, problem-solving, and authenticity to realign culture with purpose. A nonprofit director—balancing mission and resource scarcity—draws from emotional intelligence, curiosity, and motivation to sustain engagement over time. The visible expressions of competence differ, but beneath them lies the same architecture of aligned skills working in concert.

This interdependence is what makes leadership competence so difficult to develop and so valuable when it is achieved. It cannot be taught solely in classrooms or assessed through checklists. It is forged in motion, tested through tension, and strengthened through reflection. Leaders build competence by applying their skills in conditions where priorities compete and trade-offs cannot be deferred. They learn through collision: between ambition and constraint, between direction and ambiguity, between aspiration and accountability. Those collisions teach rhythm, the ability to make decisions, communicate intent, and sustain energy without waiting for clarity to arrive first.

Developing competence is, therefore, not a linear process. It does not begin and end with training programs or performance cycles. It grows through iteration across multiple contexts and through the discipline of reflection that converts experience into judgment. Skill building remains foundational, but context gives those skills meaning, and integration turns them into momentum. Competence emerges when leaders connect learning across domains, when professional, personal, and technological abilities reinforce one another instead of competing for attention.

Modern development must be designed around this reality. The most effective organizations expose leaders to complexity early, giving them responsibility across functions, geographies, and technologies. They understand that leadership readiness is built through diversity of challenge, not length of tenure. Every new context expands the leader's capacity to adapt skills into new forms, sharpening the ability to transfer judgment from one environment to another. Competence matures when leaders can absorb disruption without losing coherence, when they can interpret unfamiliar signals through the lens of shared purpose.

Competence in context is not about knowing everything; it is about knowing how to move through uncertainty without losing alignment. It is the posture that allows leaders to guide others when conditions change faster than plans can adapt. It is what separates leadership that reacts from leadership that endures. The leaders who sustain advantage are those who continue to expand and realign their skill sets, translating awareness into action in every environment they face. They create coherence in systems that refuse to stand still, proving that leadership today is not defined by experience alone but by the capacity to evolve in motion.

Leadership as the Catalyst

Leadership is not simply one variable among three. *It is the force that activates the equation of modern advantage.* Without leadership, strategy remains theoretical and technology becomes potential without purpose. Only when leadership brings credibility, clarity, and adaptability to the system do the other variables gain traction and move in rhythm. This activating role defines leadership's place at the center of modern competitive advantage.

Credibility is where the process begins. Strategy, no matter how sophisticated, will not hold if teams lack confidence in those who set direction. Technology will not achieve its potential if the people deploying it question whether it aligns with the organization's values or priorities. Credibility does not come from position or charisma; it comes from consistency. Leaders build credibility when their decisions reflect the realities of the market, when they communicate transparently, and when their actions demonstrate alignment between purpose and performance.

Clarity is the second essential quality. Modern markets are noisy—filled with data, opinions, and shifting expectations. Without clarity, organizations scatter their energy. Leadership provides the focus that allows strategy to choose and technology to prioritize. It transforms competing options into a shared path forward. Clarity is not the absence of ambiguity; it is the discipline of making direction visible and actionable, even when conditions remain uncertain.

Adaptability completes the triad. In earlier eras, strategic direction could hold steady for years, giving organizations time to adjust. Today, conditions evolve too quickly for rigidity to survive. Leadership adaptability ensures that the organization can respond without losing coherence, shifting emphasis or pace while maintaining purpose. Adaptability is not about reaction; it is about rhythm. It allows leaders to refine direction as new information emerges while keeping strategy and technology synchronized around a stable core.

When credibility, clarity, and adaptability coexist, leadership transforms from a function into an activating force. It connects ideas to execution, aligning human judgment with technological capability and strategic design. This is what allows modern organizations to move decisively even under pressure. Effective leadership converts energy into focus, complexity into coherence, and motion into traction.

Consider the example of Netflix. When streaming technology emerged, the company's leadership did not cling to its DVD-by-mail model. It acted with credibility, acknowledging openly that the existing business would not last. It provided clarity by articulating a vision for on-demand entertainment and backed that vision with investment and communication that aligned the organization. It demonstrated adaptability by iterating the model in real time, learning through audience behavior and platform data. The strategy and technology that followed were only effective because leadership created the conditions for their alignment. Without that connection, Netflix's early technology advantage would have faded into noise.

The absence of effective leadership is equally instructive. At Boeing, leadership credibility eroded as quality and safety issues surfaced. Strategy documents and technological sophistication remained, but they no longer moved together. Without leaders who could reconcile direction, culture, and engineering, the organization's coherence fractured. The result was not simply a strategic failure but a breakdown in rhythm; the system lost its pace because leadership lost its connection to the forces it was meant to harmonize.

This role does not require heroic personality or absolute certainty. It requires presence, judgment, and the ability to create conditions where strategy and technology can work together. Modern leadership is not about control but about coordination. It ensures that decisions, actions, and communication remain synchronized as conditions shift. It is the discipline of creating rhythm across systems that naturally fall out of sync.

This role is vital because leadership failures cannot be offset by excellence in the other variables. A brilliant strategy will collapse if leadership fails to inspire commitment. Advanced technology will stall if leadership fails to define purpose. Even when all three elements exist, without leadership to connect them, progress fragments and energy dissipates. Leadership is the multiplier that turns alignment into momentum.

The reflection for modern leaders is straightforward. Are you activating the system or merely occupying a position? Are you earning credibility through consistency or relying on authority for compliance? Are you creating clarity that unifies action or generating noise through competing priorities? Are you demonstrating adaptability that keeps pace with your environment or defaulting to the comfort of what once worked? The answers to these questions reveal whether leadership is enabling modern advantage or quietly undermining it.

Leadership as catalyst is not a metaphor. It is a discipline that determines whether strategy and technology achieve coherence. Credibility builds trust. Clarity channels energy. Adaptability sustains motion. When these forces operate together, leadership activates the equation of modern advantage and transforms potential into progress.

Meridiem: A Fractured Leadership Team

By the time the equation had broken, the fractures at Meridiem were visible at the top. The leadership team had long been respected. Clients valued their reliability, employees trusted their professionalism, and the board praised their discipline. For years, credibility had been their greatest strength. Yet as the market shifted and competitors adapted faster, that credibility began to erode. Respect for the individuals around the table no longer translated into clarity of direction or unity of action.

The first signs appeared in executive meetings. The chief executive still opened each session by emphasizing stability, while several direct reports called for reinvention. One argued for aggressive innovation in new products, another for deeper client relationships, and a third for faster digital integration. Each position had merit, but together, they revealed a deeper issue. The team could debate vision but not translate it into a shared plan. They discussed options in detail but struggled to decide which to pursue. Clarity turned into conversation, and progress slowed.

Communication followed the same pattern. Executives spoke often but rarely in alignment. Messages that reached middle managers were polished but inconsistent. One leader described the company as evolving cautiously; another promised transformation. Teams received both messages and responded with hesitation. Communication, once a hallmark of Meridiem's culture, had lost precision. The right information moved slowly, and purpose blurred in the noise.

Decision-making became cautious. Leaders delayed choices until consensus formed, even when the evidence was clear. The team mistook caution for discipline. Analysis replaced action, and opportunities passed quietly to faster competitors. Meridiem's culture of thoroughness, once a differentiator, had turned into drag. Evaluation overtook execution.

Adaptability was tested when a major client introduced a new procurement model that favored integrated digital solutions. Competitors responded quickly, bundling services and platforms into single offers. Meridiem hesitated. Proposals were drafted and redrafted, but no decision followed. Leadership waited for clarity that never came. By the time the team agreed to act, the opportunity had passed and the client had chosen another provider. What looked like prudence was paralysis.

Tension deepened as pressure grew. Executives who once collaborated smoothly began defending their divisions. Conversations that had once produced alignment now ended in fatigue. Employees sensed the fracture at the top and mirrored it below. The organization remained professional and polite, but the energy of purpose was gone. Meetings became routines of reassurance rather than instruments of progress.

Technology exposed the divide. The CIO pressed for more integration across platforms and functions, but other executives viewed technology as a cost center rather than a driver of advantage. Digital initiatives multiplied but remained disconnected. Systems advanced, yet decision-making stayed slow. Tools changed faster than behavior. The team could describe modernization but could not model it.

By midyear, the fractures were visible in results. Projects ran longer, client satisfaction declined, and turnover among younger employees climbed. The board questioned whether leadership still had the range to guide the company forward. In one meeting, a director asked what Meridiem wanted to be known for now. No one answered immediately. Each executive could describe their function in detail but struggled to express a coherent vision for the whole. Visioning, once second nature to the team, had faded into absence.

At the next all-hands meeting, the chief executive spoke confidently about steady progress and strategic patience. Employees listened respectfully but without conviction. When she finished, a young manager asked whether the company planned to expand its digital capabilities or partner with external providers. The question was simple. The answer was vague. Leadership had communication but not clarity, awareness but not alignment.

In the weeks that followed, several executives admitted privately that something deeper was wrong. They were managing activity rather than leading coherence. The team's capabilities reflected experience in a different era; many of the modern leadership skills the environment now demanded—such as clarity under ambiguity, adaptability across pace, and the ability to lead through technology rather than around it—had not yet been built. Their credibility remained intact, but their rhythm had broken. Leadership, the catalyst of advantage, had become the bottleneck. Until Meridiem's leaders developed the modern competencies required for coherence and reconnected their strengths into shared motion, the organization would continue to work hard without moving forward.

Chapter Summary

Modern leadership begins with capability but matures through combination. Essential skills such as communication, decision-making, adaptability, problem-solving, and emotional intelligence form the foundation of every leader's effectiveness. These skills provide structure and credibility, yet they create advantage only when they operate together as integrated competencies. Competence arises when judgment, empathy, and clarity converge under pressure, turning individual ability into collective rhythm that holds through uncertainty.

Context determines how that rhythm is applied. In every environment, whether healthcare, technology, finance, or manufacturing, the effective leader adapts skill and style to sustain coherence amid complexity. Leadership readiness is no longer measured by tenure or scope of authority but by the capacity to translate change into alignment, to guide people and technology in motion, and to renew confidence as conditions evolve. Competence in context is what transforms experience into relevance.

When these capabilities align, leadership acts as the catalyst. It activates the full equation of modern advantage, connecting strategy and technology into a single pattern of clarity, coherence, and traction. Without that activation, vision fragments and progress stalls. With it, organizations move deliberately and stay aligned as the environment accelerates. The next chapter, *Differentiated Strategy: The Second Variable of Modern Advantage*, explores how clarity of direction becomes the blueprint that channels this leadership energy into distinction and sustained advantage.

CHAPTER 5: DIFFERENTIATED STRATEGY: THE SECOND VARIABLE OF MODERN ADVANTAGE

The Blueprint for Advantage in a Modern World

Of the three variables that create modern advantage, strategy is the one most often over-defined and under-applied. Boardrooms devote hours to refining it, consultants are hired to validate it, and presentations overflow with evidence of rigor. Yet despite the attention, most organizations still treat strategy as a static artifact rather than a living system that connects leadership intent with technological capability. The result is structure without motion, plans without distinction..

The truth is that differentiation has become harder to achieve and easier to lose. In hypercompetitive markets, advantages built on cost, distribution, or brand can be replicated quickly. Customers expect more, competitors move faster, and investors reward distinction rather than imitation. A strategy that does not set an organization apart in meaningful ways is not a strategy; it is survival disguised as intent. Modern strategy must create coherence that holds as conditions change and distinction that endures as competitors imitate. It depends on leadership to activate direction and on technology to accelerate delivery, translating purpose into progress.

This is why strategy is the second variable of modern advantage. Leadership may be the catalyst that activates the equation, but strategy is the blueprint that provides coherence and direction. Without it, leadership becomes inspiration without focus and technology becomes noise without purpose. A differentiated strategy is not about being incrementally better; it is about being distinct in ways that matter and defensible in ways that endure, then applying that distinctiveness through adaptation to the realities of today's markets.

Differentiation Is Cumulative, Not a Silver Bullet

The concept of a silver bullet is the greatest threat to strategic differentiation. Leaders often assume that one decisive move, a new product, market, acquisition, or technology, will set their organization apart. These gestures appear bold and decisive, and they appeal to the desire for simplicity in complex markets. Yet differentiation is rarely created in a single moment. True distinction is built through connection, through reinforcing choices that operate in rhythm with one another and with the broader system of leadership and technology. It is the excellence within strategy and the synchronization across the variables that make advantage distinct, durable, and hard to replicate.

The silver-bullet mindset does more than misread competition; it misreads how advantage is created. In hypercompetitive markets, even significant moves are quickly matched. A product innovation may create a brief surge in attention, but rivals soon respond. A geographic expansion may open new opportunities, but incumbents adapt. A cost initiative may improve margins, but competitors can imitate it. None of these, by themselves, create defensible advantage. They generate temporary results. Modern differentiation arises when these moves are connected, layered, and reinforced by a coherent blueprint; competitive advantage emerges when that distinctiveness is synchronized with leadership and technology.

Consider Peloton. Its innovation was compelling, and for a time it appeared to have achieved a breakthrough. However, its position was not reinforced by other elements of differentiation. Leadership was divided, its broader strategy unclear, and its technology confined to a single context. What looked like a silver bullet proved fragile. When demand normalized, Peloton had no cumulative approach to sustain momentum.

Apple offers the opposite story. Its advantage has never rested on one product alone. Each innovation deepened a coherent ecosystem of hardware, software, and services. That system created stickiness that competitors could not easily replicate. Apple's strategy was cumulative, each move reinforcing the next and contributing to an advantage that endured even as individual features were copied.

The same principle applies far beyond technology. A mid-sized logistics company that adopts route optimization tools may see short-term efficiency gains. But when leadership aligns around a digital-first strategy, retrains staff to deliver on that promise, and embeds technology into client processes, the results multiply. Customers experience speed and reliability. Employees see focus. Investors recognize momentum. Advantage is no longer about a single tool; it emerges from excellence within strategy and synchronization across variables, with differentiation operating as the connective pattern others find difficult to match.

Leaders must resist the illusion of the silver bullet. It is compelling because it offers the fantasy of progress without the discipline of coherence. Yet sustainable advantage cannot be achieved in a single stroke. It must be designed as a living system, a strategy in motion in which excellence within its own design reinforces synchronization across the broader equation. The reflection for leaders is demanding: Are your strategies connected by a coherent blueprint, or are they isolated bursts of activity? Are you designing for continuous reinforcement or chasing momentum that fades?

In today's marketplace, organizations that win are not those that chase the biggest headlines. They build cumulative systems of differentiation, where every choice reinforces the others, where leadership and technology move in step with the blueprint, and where no single move carries the weight of sustaining advantage. Differentiation sets the conditions; advantage follows when those conditions hold in execution.

True Differentiation Requires Distinctiveness

In many organizations, strategy discussions drift toward incremental improvement. Leaders talk about being better than competitors, offering faster service, lower costs, higher quality, or more features. These goals are not wrong, but they are insufficient. In markets where advantage is fragile, simply being better rarely creates lasting distinction. Customers, investors, and stakeholders reward organizations that are not just incrementally better but meaningfully different. Distinctiveness, not marginal improvement, is what defines modern strategy.

The comfort of incremental goals lies in their safety. They are easy to measure, easy to communicate, and easy to defend. Leaders can point to efficiency gains, improved satisfaction, or modest share growth as evidence of progress. Yet these improvements are easy for competitors to match. A faster process can be replicated. A lower price can be matched. A quality upgrade can be copied. Incremental improvements reset the baseline of competition rather than create advantage.

Distinctiveness, by contrast, is harder to imitate. It demands choices that separate an organization from peers in ways that matter to the market. Distinctiveness may come from a unique business model, a differentiated customer experience, or a strategic focus that competitors overlook. It is not about doing everything better; it is about doing certain things differently, in ways customers recognize and competitors struggle to replicate.

Consider Tesla. The company did not simply build cars with better fuel efficiency; it reframed what a car could represent. Performance, design, and software worked together to redefine the category. That distinctiveness made Tesla a symbol of innovation, even as other automakers entered the electric market. Its position may now be contested, but its early distinctiveness shifted the entire industry. Tesla was not better in a conventional sense; it was different in a way that mattered.

Service industries show similar dynamics. When Southwest Airlines committed to low-cost, point-to-point flights, it was not just trying to be cheaper. It redesigned the experience, simplified operations, standardized turnaround times, and built a service culture that competitors could not easily mimic. For decades, that distinctiveness set it apart, even as larger airlines copied pieces of the model. The advantage was not marginal improvement; it was a different logic of competition.

Smaller organizations can achieve the same. Canva entered a crowded market for professional design software and chose not to compete on advanced features. It focused on simplicity and accessibility, opening design to non-experts. Canva's differentiation was not in being a better Photoshop; it was in offering a different kind of design tool, tuned to unmet needs and executed with coherence.

Distinctiveness requires courage. It means making deliberate trade-offs, saying no to markets, features, or practices that dilute focus. These choices often attract criticism as stakeholders question why the organization is not pursuing what others are doing. Yet distinctiveness is forged in those moments. It endures only when leadership and technology remain synchronized behind it, when direction, design, and delivery reinforce one another in motion.

Leaders should ask themselves: If our name were removed from our strategy documents, would the plan still look different, or could it be mistaken for any of our competitors? Distinctive strategies pass this test. They are recognizable even without the logo. They tell a story only one organization can credibly deliver.

Achieving distinctiveness demands conviction. Many organizations aim for superiority across too many dimensions, spreading resources thin. The result is competence without character, good at many things, distinctive in none. Customers cannot articulate why they should choose it, competitors can match its moves, and investors see effort without momentum. Distinctiveness requires the discipline to focus, the judgment to choose, and the patience to stay consistent when imitation arrives.

In the modern environment, distinction is a primary currency within strategy and a core ingredient of advantage. It attracts customers, wins talent, and earns investor confidence. Better is temporary; different is defensible. Distinctiveness captures attention; stickiness keeps it. Distinctiveness sets the stage; advantage is realized when that distinctiveness is sustained in execution and reinforced by technology.

Stickiness Makes Strategy Durable

Differentiation alone is not enough. A strategy may set an organization apart for a time, but if its distinctiveness can be copied or replaced quickly, the advantage will not last. In modern markets, where ideas and features spread instantly, the true test of strategy is whether it is sticky, whether it embeds value in ways that are difficult for competitors to replicate or for customers to abandon.

Stickiness comes from more than customer satisfaction. Customers can be satisfied and still switch when a cheaper or faster alternative appears. Stickiness emerges when excellence within each variable connects through synchronization across them, when credibility, coherence, and capability reinforce one another in motion. It is the system effect that makes advantage both visible and durable.

Technology companies often illustrate the idea. Apple and Netflix built ecosystems that link products, data, and experiences so tightly that customers think twice before leaving. Competitors can copy features, but they cannot easily reproduce the web of connections that create continuity and convenience. Stickiness in these cases is not the result of a single product choice; it is the outcome of design, rhythm, and reinforcement across leadership, strategy, and technology, which converts distinctiveness into durable advantage.

Other industries demonstrate the point in different ways. Some banks have created stickiness by embedding themselves in clients' daily operations. Business banking platforms that connect payments, payroll, and analytics are not easily replaced. Even if a rival offers better pricing, the cost and complexity of switching become deterrents. The strategy is defensible not because it is always the cheapest or most innovative but because it is embedded. Professional services firms show another dimension. Their relationships compound through trust and institutional knowledge, making replacement risky and slow. Brands such as Patagonia show that stickiness can be emotional as well as structural, where customers stay because the relationship reflects their values.

Stickiness is not only external. It exists inside the organization as well. When a strategy is coherent, people know how to make trade-offs without escalation, handoffs are cleaner, and capital flows to what matters most. That internal stickiness, the habit of acting in rhythm, turns direction into everyday discipline. It shortens re-decisions, reduces rework, and makes renewal easier because decisions compound rather than collide.

The absence of stickiness explains why many seemingly strong strategies erode quickly. A company may launch a new app with innovative features, but if those features are easy to copy, users migrate as soon as another option appears. A retailer may differentiate with a trend or campaign, but if nothing binds customers beyond the moment, loyalty fades. Without stickiness, differentiation becomes temporary.

Leaders must design for stickiness deliberately. They should ask how difficult it is for customers or partners to leave, what they lose if they switch, and what it would take for a competitor to replicate not one feature but the full experience the organization provides. These questions shift the mindset from being different in the moment to being defensible over time.

Stickiness also requires vigilance. What feels defensible today may not remain so tomorrow. Ecosystems can be disrupted, relationships can be disintermediated, and data advantages can erode. Leaders must continually reinforce stickiness by evolving how value is delivered and by deepening the connections that make leaving unattractive. Stickiness keeps advantage durable, and embracing modern realities keeps it relevant.

Differentiation Must Embrace Modern Realities

Many strategies fail not because leaders aim too low, but because they build on foundations that no longer hold. For much of the last century, organizations created advantage through scale, distribution, or brand. Those forces still matter, but they no longer guarantee relevance. In today's environment, differentiation must be designed for speed, convergence, and complexity— conditions that redefine how advantage is created and how long it lasts.

Modern differentiation depends on how strategy behaves in motion. It can no longer stand apart while leadership and technology evolve around it. Strategy must translate rapid change into coherence and direction, keeping purpose clear even as context shifts. When strategy loses that rhythm, even well-crafted plans drift out of sync with the markets they serve.

The trap for many leaders is treating yesterday's playbook as if it will hold tomorrow. They assume that defending market share, sustaining brand reputation, or protecting scale will keep them safe. Yet markets now move faster than structures can adjust. Scale can be undercut by agility. Distribution can be bypassed by digital channels. Brand loyalty can evaporate when customers find more personal alternatives. Legacy strengths provide temporary comfort, not lasting differentiation, unless they are reframed within modern dynamics.

Retail and financial services show this clearly. Large chains once relied on store networks or branch footprints as their moat. That position held until digital-first entrants reshaped expectations around access and transparency. Incumbents that clung to physical scale fell behind, while those that redefined their strategies around digital connection, last-mile delivery, and data coherence created new forms of advantage. The same pattern appears in healthcare, where telemedicine and AI-enabled diagnostics have moved from novelty to necessity. Providers that still compete on size or prestige are losing to those that align culture, strategy, and technology around access, speed, and experience.

Technology itself is a double-edged force. Cloud platforms, AI, and digital infrastructure have lowered barriers to entry, making features easier to copy. Yet they also open space for organizations that can embed technology inside a differentiated strategy, turning common tools into uncommon experiences. Shopify, for example, did not lead with the most advanced e-commerce features. It differentiated by connecting technology with a strategy to empower entrepreneurs, establishing a position competitors could not easily reproduce and strengthening its path to advantage.

Stakeholders amplify this shift. Investors now expect evidence of sustainability and governance. Customers demand values as well as value. Employees choose employers for culture and purpose, not just pay. A strategy that ignores these dimensions quickly loses credibility. Differentiation today must account for this broader field of expectations, recognizing that stakeholders judge coherence across every promise an organization makes.

Modern advantage also requires a faster loop between sensing and shaping. Strategy can no longer be a once-a-year ritual; it must function as an operating rhythm that continually interprets signals from markets, technology, and stakeholders. The organizations that sustain differentiation build mechanisms such as forums, analytics, and feedback loops that allow them to adjust intent without losing coherence. Agility replaces reactivity because direction and execution stay connected in real time.

Healthcare illustrates the point again. Hospitals once competed on reputation or scale, but patients now prioritize access, convenience, and digital engagement. Telehealth and remote monitoring are not side projects; they are the new front door. Leaders who continue to frame differentiation around physical capacity or historical prestige are competing on the wrong axis. Those who align culture, strategy, and technology around accessibility are building modern advantage.

Modern differentiation, therefore, requires anticipation as well as adaptation. Leaders must continually test which assumptions are fading and which new realities are shaping the basis of competition. Strategy must move in rhythm with its environment, refreshing distinctiveness before it decays. The reflection for leaders is simple but demanding: Which beliefs about our market are no longer true? Which emerging realities are redefining the rules? Are we updating our blueprint fast enough to stay credible and coherent?

Differentiation that embraces modern realities is harder to achieve but far more powerful. It creates positions competitors cannot easily undermine because they are built on the forces shaping the future rather than the past. Strategy that evolves with the world becomes the blueprint that keeps leadership and technology aligned in motion.

Strategy as the Blueprint

Leadership may activate the system, but strategy gives it structure. It is the design that connects leadership's intent with technology's capability, turning energy into direction and direction into momentum. Without it, leadership becomes inspiration without focus and technology becomes motion without purpose.

Modern strategy is not a plan on paper; it is a living architecture that shapes how choices interlock and how progress compounds. It defines the logic of coherence, how decisions in one area reinforce those in another, how investments translate intent into action, and how adaptation occurs without losing direction. The blueprint does not slow motion; it gives it form.

Too many organizations still mistake planning for strategy. Plans describe what might happen; strategy decides what must happen and what will not. Those decisions create distinction and coherence. A real blueprint focuses attention, channels resources, and ensures that every initiative strengthens the system rather than competes with it.

The blueprint matters because it turns aspiration into design. With it, leadership's intent is translated into trade-offs that technology can scale and teams can execute. Without it, energy disperses, functions optimize locally, and progress fragments. The blueprint keeps the enterprise coherent while allowing renewal to occur continuously.

When leadership provides clarity and conviction and technology amplifies those priorities through systems and data, strategy binds the motion into pattern. When either variable drifts, the design weakens and energy becomes noise. Adobe's shift to a subscription model shows what coherence in action looks like. Leadership framed the purpose, technology enabled the model, and strategy orchestrated the transition. The blueprint tied intent to execution and converted change into advantage.

A disciplined blueprint also protects focus amid noise. Every enterprise faces more opportunities than it can pursue. Without design, effort scatters; with it, leaders can allocate attention where it matters most, keeping the organization balanced even as conditions evolve.

Strategy as a living blueprint converts direction into differentiation and activity into advantage. It channels leadership's energy, anchors technology's pace, and ensures that motion compounds rather than collides. When strategy operates this way, coherence becomes capacity, and advantage renews itself in rhythm with the world.

From Blueprint to Motion

Even the best strategy can stall if it never leaves the page. Execution is where the blueprint proves its worth, where coherence is tested under pressure and synchronization becomes real. A strategy earns its credibility not through elegance in presentation but through the consistency of decisions and behaviors it produces across the system.

Executing the blueprint begins with clarity. Every person and function must understand not only what to do but why those actions matter to the organization's direction. When a blueprint is clear, decision-making accelerates because priorities are self-evident. Teams no longer wait for instruction; they act within intent. This is how leadership's energy becomes distributed without losing focus.

The second test of execution is coherence. A strategy that cannot guide trade-offs will collapse into departmental logic. The blueprint's role is to make the connections visible: how investment choices in one area enable progress in another, how technology deployments reinforce design, and how cultural signals sustain momentum. When decisions align with that architecture, execution feels faster not because people work harder but because the system moves as one.

Consider how strategy execution changed at Adobe once the company adopted its subscription model. The blueprint was clear: recurring value through continuous delivery. Every decision, from pricing to customer communication to data infrastructure, was filtered through that lens. Technology enabled the rhythm, but strategy set the tempo. The coherence between the two created not only efficiency but cultural traction, a shared understanding of what progress looked like.

Execution also exposes where synchronization fails. When technology evolves faster than leadership decisions, coherence frays. When leadership communicates intent but ignores operational reality, momentum dissipates. Successful organizations build feedback loops between vision and delivery so information flows in both directions. The blueprint provides direction, and execution refines it. Together, they form a living cycle that keeps strategy relevant and leadership credible.

The reflection for leaders is practical. Does your strategy translate clearly into daily action? Can every team see how their work reinforces the larger design? Are your feedback systems fast enough to adapt without losing coherence? In modern competition, advantage depends not only on having a strategy but on executing one in rhythm with leadership and technology.

When a blueprint functions in motion, execution ceases to be a separate discipline. It becomes the visible expression of synchronization across the variables. That is what turns intent into impact and strategy in design into advantage in action.

Meridiem: The Cost of an Undefined Blueprint

By the time Meridiem's leadership gathered for its strategy reset, the cracks were clear. Growth had slowed, client retention was slipping, and competitors were winning with sharper positions in the market. The team agreed that change was necessary. Yet over two days of presentations and debate, it became clear that Meridiem did not have a strategy that could set it apart. It had activity, initiatives, and ambition, but no blueprint for advantage.

The proposals sounded impressive. The product group argued for expanding the portfolio, convinced that more variety would attract clients. The client services leaders pushed for deeper relationships, relying on Meridiem's reputation for reliability. The technology team pressed for faster digital investment to close the gap with competitors. Each argument was rational on its own, but together they revealed the absence of a unifying design. Leadership, strategy, and technology were no longer moving in rhythm, and their motion lacked coherence.

That fragmentation appeared quickly in the market. In one major client pursuit, teams received conflicting guidance from executives. One emphasized reliability, another innovation, and a third transparency enabled by new tools. The result was confusion. The proposal tried to do everything and convinced no one. A long-standing client described it as safe but forgettable before awarding the work to a competitor that promised sharper value, with speed and clarity supported by connected systems. Meridiem's reputation for reliability no longer differentiated; it simply blended into the background.

Inside the company, fatigue deepened. Managers were asked to execute on multiple priorities at once—including service excellence, digital pilots, and cost control—without clear hierarchy or focus. "We are chasing everything," one manager admitted, "and standing still at the same time." The lack of coherence drained energy at every level. Younger employees, eager to modernize, hesitated without direction from the top. The organization moved, but not together.

The board noticed the drift. In quarterly reviews, directors asked pointed questions. What makes Meridiem different? How do investments in technology connect to strategy? Why are client win rates falling despite strong individual capabilities? Leadership responded with orderly slides, but the coherence behind them was gone.

Investor pressure followed. On the next earnings call, analysts asked what Meridiem's differentiator was. Executives spoke of reputation and reliability, but those were echoes of the past, not promises for the future. Within weeks, the stock slipped. The market recognized what clients and employees already knew: Meridiem was credible but indistinct.

The consequences compounded. Competitors with coherent strategies pulled ahead. Employees lost faith in leadership's direction, and the board grew impatient. Meridiem's failure was not a lack of talent or effort but the absence of a working blueprint, the mechanism that keeps leadership, strategy, and technology synchronized in motion. Respect for individuals could not substitute for organizational coherence.

Meridiem's story shows what happens when credibility outlasts clarity. Without a blueprint that aligns choices, channels focus, and connects technology to direction, activity becomes noise. Until its leaders rebuild the architecture that links intent to action, the modern approach will remain out of reach and the organization will continue to work hard without moving forward.

Chapter Summary

Silver bullets collapse, incremental gains fade, and legacy strengths erode, leaving only those with a coherent blueprint to stand apart. Modern competitive advantage demands strategy that is cumulative, distinctive, sticky, and reflective of modern realities. It is through this discipline that differentiation becomes both deliberate and defensible, and that distinctiveness contributes to competitive advantage. Sustaining it requires endurance as much as ingenuity. Strategies that hold are treated as processes, not proclamations, and systems that renew themselves through feedback and synchronization.

Leadership may catalyze action, but strategy provides coherence. It is the blueprint that turns direction into coherent choices and those choices into differentiation that endures. Without it, leadership becomes inspiration without focus, and technology becomes motion without purpose. With it, leadership gains traction and technology gains meaning. Together, they form a synchronized system capable of sustaining advantage even as conditions change.

The reflection for leaders is clear. Does your strategy set you apart, or does it sound like everyone else's? Does it build stickiness that sustains advantage, or is it easily copied? Does it adapt to modern realities, or does it cling to assumptions that no longer hold? Modern competitive advantage belongs to those who can answer these questions honestly and act decisively. The next chapter, *Integrated Technology: The Third Variable of Modern Advantage,* explores how embedding technology within leadership and strategy turns clarity into traction and makes advantage real in the marketplace.

CHAPTER 6: INTEGRATED TECHNOLOGY: THE THIRD VARIABLE OF MODERN ADVANTAGE

The Enabler of Modern Advantage

Technology is the most dynamic of the three variables of modern advantage, yet it is often misjudged in how it truly creates value. Organizations invest heavily in new platforms, tools, and pilots, often convinced that the next upgrade will deliver an edge. Boards celebrate major IT initiatives, and leaders point to digital dashboards or AI programs as proof of progress. Yet despite these efforts, few organizations translate technology spending into sustainable advantage.

The reason is straightforward: Technology alone does not create advantage. Most competitors have access to similar tools, and features that seem advanced today are copied tomorrow. What separates leaders from also-rans is not ownership of technology but the ability to integrate it. When technology is embedded in strategy and reinforced by leadership, it becomes the mechanism that gives the equation traction. When it operates in isolation, it consumes resources without creating distinction. Integration is what turns technology from a collection of tools into a capability that advances performance.

Modern advantage demands more from technology than technical soundness. Excellence inside the domain still matters, but its real value emerges through connection. Technology must scale reliably, adapt quickly, and link seamlessly with leadership's priorities and strategy's intent. The organizations that excel treat integration as design, not coordination. Their investments reinforce purpose, accelerate execution, and sustain pace. That is how technology turns potential into performance and spending into advantage.

This is why technology stands as the third variable of modern advantage and why the term "integrated" matters. Leadership and strategy are recognized as enterprise-wide disciplines; *technology must now be treated with the same breadth*. In many organizations, it remains confined to specialists or isolated functions, a department to be managed rather than a source of coherence. In the modern environment, that separation is costly.

Technology's role is distinct. It does not catalyze like leadership or provide the blueprint like strategy. It enables. It is the force that converts credibility and clarity into motion, allowing organizations to deliver at the pace and scale demanded by modern markets. Without integrated technology, leadership and strategy remain aspirations. With it, they become action. This is why technology completes the equation and serves as the enabler of modern advantage.

Technology Is Foundational

In earlier eras, technology was often treated as a support function. It automated processes, managed infrastructure, and improved efficiency, but it was not seen as central to competitive strategy. That view no longer holds. In the modern environment, technology is not an accessory; it is foundational. It reshapes industries, redefines how organizations deliver value, and alters the very conditions of competition. Without integrated technology, leadership and strategy cannot gain traction.

The evidence is visible across industries. In healthcare, the expansion of telemedicine and AI-driven diagnostics has not merely improved efficiency; it has redefined how patients access care and how providers organize their operations. In retail, predictive analytics and digital platforms have transformed customer expectations around personalization and fulfillment. In manufacturing, automation and advanced data analytics have enabled new business models centered on predictive maintenance and service-based offerings. In every case, technology has changed the nature of competition itself.

What makes technology foundational is not only its breadth but its pace. Capabilities that once offered an edge quickly become the norm. A digital service may appear innovative on launch, but competitors can match it within months. Cloud infrastructure, AI tools, and digital platforms are widely available, lowering barriers to entry. Organizations can no longer rely on first-mover advantage alone. They must embed technology across the operating model so that it reinforces leadership clarity and strategic distinctiveness. Embedding transforms capability into capacity and makes technology a true foundation of advantage.

Consider Shopify. Its leadership positioned technology as the backbone of an ecosystem, not a feature set. The company's platform unified e-commerce, payments, and logistics for millions of businesses while enabling an open developer network that extended its reach. Shopify's strength did not lie in any single capability, but in the way its technology was embedded as infrastructure: supporting leadership's mission to empower entrepreneurs and anchoring a strategy built on accessibility and scalability.

Now contrast that with Peloton. For a time, its connected hardware and digital content appeared revolutionary. The technology was compelling, and the user experience inspired loyalty. Yet the company struggled to turn that momentum into sustainable advantage. Its technology remained excellent, but the organization treated it as a product line rather than a foundation. When competitors introduced comparable offerings, Peloton's distinctiveness faded. Its experience illustrates how quickly technology loses strategic weight when it is not embedded in leadership vision or operational coherence.

Smaller organizations demonstrate the same principle. Enel, the Italian energy provider, once relied on conventional grid operations. By investing in smart infrastructure and embedding digital systems into the core of its business, Enel shifted from a traditional utility to a data-driven energy services provider. Leadership tied the transformation directly to its sustainability strategy, making technology a foundation for both environmental impact and shareholder value. The advantage emerged not from tools but from coherence across leadership, strategy, and technology.

This is the shift leaders must internalize. Technology is no longer something added to reinforce advantage; it is the foundation on which advantage is built. Without it, leadership and strategy lose traction. A strategy that is not digitally fluent is obsolete before it launches. Leadership that avoids or misjudges technology loses credibility before it gains momentum. Technology, leadership, and strategy must move together, and technology often determines how quickly advantage can take hold and endure.

Modern advantage depends on seeing technology not as an accessory but as a foundation. It reshapes industries, resets expectations, and accelerates cycles. It is what allows leadership and strategy to gain traction and without it, even the most credible leaders and distinctive strategies will falter.

Technology Excellence Defines the Standard

In many organizations, technology is still treated as a secondary concern. Leaders speak passionately about vision and strategy, but when it comes to technology, they defer to specialists or view it as a cost center. This separation reflects a lingering assumption that leadership and strategy drive performance while technology merely supports it. That assumption no longer holds. In the modern environment, excellence in technology is as critical as excellence in leadership or strategy. Without it, the organization cannot create or sustain advantage.

The reason is straightforward. Strategy cannot succeed without the means to execute, and in today's market, execution depends on technology. A brilliant blueprint may set a differentiated course, but without integrated technology, it remains theory. Similarly, leadership may inspire credibility and align teams, but without technology to enable delivery, those commitments collapse under the weight of reality. Technology is no longer a back-office function; it is the mechanism by which strategies come to life and leadership signals gain traction. Excellence within technology keeps the system synchronized, translating direction and credibility into execution that holds under pressure.

Organizations that underestimate this truth quickly discover the consequences. Capital One provides an instructive example. Its leadership built a strategy around personalization and data-driven decision-making, but early progress stalled until the company committed fully to cloud transformation. By embedding technology at the center of its operating model, Capital One unified data, reduced complexity, and improved agility. The outcome was not simply efficiency but renewed differentiation: proof that technology excellence converts strategy from plan to performance.

The same pattern appears in industries defined by speed. Ocado, the U.K.-based online grocer, redefined supply-chain performance by treating technology as its primary product. Leadership aligned the business around robotics, automation, and analytics, building an operating model in which software, logistics, and strategy worked as one system. Competitors with similar tools could not match its execution discipline or precision. Ocado demonstrates that technology excellence, applied with clarity, turns scale into advantage.

Excellence also drives resilience in creative industries. Adobe's integration of Firefly, its generative-AI suite, exemplifies how technology embedded in strategy sustains relevance. Leadership framed AI not as disruption but as amplification, integrating it directly into Creative Cloud's core products. The result was not novelty but depth: a continuous feedback loop between users, data, and design that reinforced Adobe's market position and validated its leadership credibility.

In Asia, Ping An shows how technology excellence can redefine scale. Once a conventional insurer, it evolved into a technology-first enterprise through disciplined digital transformation. Leadership invested in platforms that connected health, finance, and customer engagement into one seamless ecosystem. The company's strategy became inseparable from its technology foundation, generating both efficiency and market distinction.

Excellence in technology does not mean chasing every new tool or overinvesting in novelty. It means ensuring that technology choices are disciplined, prioritized, and aligned with the organization's blueprint. It means avoiding scattered pilots and instead building systems that reinforce strategic intent. Excellence is measured not by spend or sophistication but by whether technology creates traction in the market and coherence across the organization.

Leaders must confront a difficult reality. Neglecting technology is no different from neglecting leadership or strategy. An organization with poor leadership loses credibility. One with weak strategy loses coherence. One with inadequate technology loses traction. In all cases, advantage erodes. Excellence in all three variables is required because weakness in any one undermines the whole.

Together, they form a single system of performance. In the modern environment, there is no hierarchy among the variables.

Leadership catalyzes, strategy provides the blueprint, and technology enables. Each must achieve excellence because weakness in any one drags the others down. Technology is not the support act; it is a co-equal player and, increasingly, the force that determines how fast leadership and strategy gain traction.

Technology Fluency Is Now Essential

Executives often hesitate when conversations turn to technology. They nod at presentations, approve budgets, and sponsor initiatives, but many still regard technology as a domain for specialists. They assume their role is to endorse, not to engage. In the past, that distance was manageable. Technology supported the business but rarely shaped its trajectory. In the modern environment, that distance is dangerous. Leaders do not need to be technical experts, but they must be technology-fluent. The ability to understand, question, and prioritize technology has become one of the defining capabilities of modern advantage.

Fluency is not about coding or engineering; it is about comprehension and judgment. A technology-fluent leader understands how emerging tools alter the basis of competition. They can distinguish between hype and substance. They know how to ask the right questions: What business problem does this tool solve? How does it connect with our strategy? What investments will it require, and what risks will it introduce? They can evaluate trade-offs not in technical detail but in business terms. Fluency keeps technology decisions aligned with leadership and strategy, ensuring that the organization moves as one.

The absence of fluency shows up quickly. Boards are filled with directors who have financial and operational expertise but limited technology understanding. As a result, they rely heavily on specialists, approving initiatives they do not fully grasp or hesitating to approve the ones they should. The organization either overspends on novelty or falls behind through caution. The gap is not in technology itself but in leadership comprehension.

Consider Magic Leap, the company once heralded as the future of augmented reality. It secured billions in funding and attracted world-class engineers, yet its leadership struggled to articulate how the technology would translate into real market value. Vision and invention were abundant, but strategic fluency was missing. Leaders could not connect product potential to business context or customer readiness. The technology was brilliant, but the lack of leadership fluency turned possibility into drift.

Contrast that with Adobe's integration of AI into Creative Cloud, which shows how leadership fluency shapes direction. Executives did not need to understand every algorithmic nuance; they needed to understand implications. They asked how generative AI could amplify their users' creativity, accelerate workflows, and extend Adobe's ecosystem. Their fluency guided investment, partnerships, and messaging. As a result, Adobe turned technological potential into differentiation that competitors were forced to follow.

Fluency is equally critical in industrial settings. John Deere provides a powerful example. Its leadership team is not composed of technologists, yet it has embraced precision agriculture, AI-driven equipment, and data platforms as core strategic levers. Executives ask business-grounded questions: How do these systems improve yield predictability, sustainability, and customer loyalty? Their fluency ensures that technology enhances operational results rather than existing as a separate agenda.

Fluency does more than improve decisions; it builds credibility with stakeholders. Employees notice when leaders understand the tools they are asked to use. Investors gain confidence when leaders can explain technology choices in business terms. Customers trust organizations whose leaders can articulate how technology enhances value. Fluency signals competence, alignment, and adaptability. In a world where advantage is fragile, those signals matter.

Leaders who lack fluency risk becoming passengers. They delegate technology choices entirely to specialists, assuming that expertise equals alignment. Yet specialists often optimize for their domains rather than for the organization as a whole. Without leadership fluency to connect technology to strategy and credibility, investments drift. Projects proliferate, but coherence is lost. Leaders must remain in the driver's seat, guiding decisions even if they are not writing the code. Fluency connects expertise to alignment, ensuring that technology choices serve strategy rather than the other way around.

Technology fluency will not guarantee success, but the absence of fluency almost guarantees failure. Leaders who cannot connect technology to strategy and credibility will find their organizations drifting, no matter how skilled their technical teams may be. Fluency does not require expertise; it requires curiosity, comprehension, and judgment. Leaders who cultivate it provide their organizations with one of the rarest and most valuable capabilities in today's market: the ability to turn technology from noise into advantage.

Integration Is What Makes Technology the Enabler

Technology is often the most heavily funded part of an organization's agenda yet the least integrated. Budgets swell with digital initiatives, AI pilots, and infrastructure upgrades. Dashboards are built, platforms launched, and apps deployed. But when these efforts remain isolated, they rarely translate into advantage. They generate activity, not traction. What makes technology the enabler of modern advantage is not its novelty or sophistication but its integration, the way it is embedded into leadership direction, strategic coherence, and daily execution.

Organizations that fail to integrate technology follow a familiar pattern. They invest in tools but treat them as projects rather than as parts of the operating model. They showcase innovation but struggle to connect it to market outcomes. Pilots multiply but seldom scale, consuming resources without creating distinction. In these cases, technology produces motion without momentum: impressive in appearance, inconsequential in impact.

Quibi provides a cautionary example. The company raised nearly $2 billion to reinvent mobile entertainment through short-form streaming. Its technology worked flawlessly, but it was never integrated into a coherent business model. The product's ambition was clear, yet leadership failed to connect it to audience behavior or distribution partnerships. When technology is detached from leadership judgment and strategic logic, even flawless execution delivers no advantage.

Asana shows a subtler version of the same issue. Its workflow platform gained a passionate user base, but its technology was not tightly embedded into enterprise ecosystems. The product excelled in functionality but lacked integration with the broader productivity architectures where decisions were made. As a result, competitors that aligned similar tools with leadership systems and strategic frameworks captured the larger share of enterprise adoption.

By contrast, Spotify demonstrates how integration turns technology into momentum. Leadership positioned technology as the connective tissue between content, data, and user experience. Streaming algorithms informed creative decisions, data guided product design, and community features shaped retention. The platform's advantage came not from a single innovation but from coherence among leadership vision, technology design, and strategic focus. Integration converted complexity into rhythm.

Traditional industries reinforce the same point. UPS's ORION platform, which optimizes delivery routes using real-time data, has become an invisible but essential driver of efficiency and sustainability. Leadership did not treat it as a technology project but as a business system that redefined performance metrics and decision-making. The outcome was not only cost savings but also a structural advantage that competitors could not easily replicate.

Smaller organizations can achieve similar outcomes. A mid-sized European apparel manufacturer integrated digital tracking into its supply chain, linking sustainability data with customer communication. Leadership tied the investment directly to brand credibility and regulatory readiness. The technology itself was simple, but its impact came from how it was embedded into both strategy and story. Integration turned compliance into trust and transparency into growth.

Integration is not just technical alignment; it is organizational coherence. Leadership must provide clarity on what technology is meant to achieve. Strategy must define where it adds distinction. Governance must ensure that investments are prioritized and scaled to reinforce direction. Without these connections, even sophisticated tools remain side projects. With them, technology accelerates advantage.

This explains why many organizations appear perpetually busy with technology but rarely advance. Leaders announce initiatives, teams build platforms, and dashboards glow with activity, yet none of it compounds because the work is fragmented. Activity without integration creates motion without progress.

Integration turns technology from potential into performance. It transforms pilots into platforms, features into systems, and data into decisions. Without integration, technology distracts. With integration, it enables leadership and strategy to take shape in the marketplace.

Governing Pace and Complexity

Technology's potential is undeniable, but so are its challenges. The pace of change is relentless, and the complexity of choices can overwhelm even capable organizations. Leaders face a flood of new tools, vendors, and platforms, each promising transformation. Employees are introduced to new systems before mastering the last. Boards are pressed to fund initiatives they cannot fully evaluate. In this environment, the risk is paralysis. Organizations either move too slowly, waiting for clarity that never comes, or chase too many initiatives at once, scattering energy without creating advantage.

Mastering pace and complexity is not about keeping up with every new tool. It is about governance: the guardrails and disciplines that ensure technology choices reinforce leadership and strategy rather than overwhelm them. Governance keeps pace synchronized with purpose, turning speed from a threat into a source of traction. The pace of innovation cannot be slowed, but it can be filtered. Leaders must build mechanisms to evaluate which technologies matter, how they connect to the organization's blueprint, and whether they can scale across the enterprise. Without this discipline, organizations either chase novelty endlessly or delay until advantage has already slipped away.

Toyota provides an enduring example. As the company expanded from hybrids to fully electric and hydrogen vehicles, leadership resisted the pressure to pursue every emerging technology simultaneously. Governance anchored innovation to a clear purpose: sustainable mobility delivered through reliability and long-term value. Every major initiative passed through the same filter: does it reinforce efficiency, quality, and the customer promise? This discipline allowed Toyota to modernize at a deliberate pace, converting technological abundance into durable progress.

Industrials offer similar lessons. Siemens Digital Industries faces constant waves of innovation in automation, sensors, and data analytics. Its governance model requires that each new investment strengthen the company's unified industrial software platform rather than fragment it. That clarity enables Siemens to innovate continuously without losing coherence, proving that complexity can be managed through alignment instead of volume.

Smaller organizations can apply the same discipline. A regional energy provider—inundated with proposals for AI forecasting, digital twins, and IoT sensors—established a simple governance rule: Every initiative must connect directly to its strategic goals of reliability, sustainability, and transparency. Projects that failed the filter were deferred. The result was not slower progress but more meaningful progress. By embedding governance into decision-making, the company converted pace from chaos into cadence.

For leaders, the reflection is practical and immediate. Are your governance systems filtering pace and complexity into coherent choices, or are they allowing technology to proliferate unchecked? Where does innovation reinforce leadership clarity and strategic intent, and where does it simply create noise? Which projects are building traction across the organization, and which are scattering energy without return? Mastering pace and complexity is not about acceleration. It is about discipline, the discipline to align speed with purpose and abundance with advantage.

Meridiem: Technology Without Traction

Meridiem's leadership believed they had finally made a bold move. After months of debate, they approved a flagship AI initiative, complete with a launch campaign designed to signal modernization. The project was framed as a turning point, proof that Meridiem was closing the technology gap with competitors.

The investment was substantial. The system promised predictive analytics, elegant dashboards, and automated reporting. Executives highlighted it in board updates, clients were invited to demonstrations, and employees were encouraged to attend town halls describing its potential. On the surface, it looked like a leap forward.

The cracks appeared quickly. The initiative was managed almost entirely by the technology group. Leadership endorsed it but did not embed it in a broader narrative about Meridiem's direction. Strategy remained unchanged, and the new system was never positioned as part of a differentiated promise to clients. It was a technology project, not an operating-model shift.

Sales teams struggled first. They showcased the system's features, but clients did not see how it solved their deeper needs. "It looks impressive," one client remarked, "but I don't understand how it changes the way you work with us." Without integration into strategy, the initiative became a demonstration, not a differentiator.

Employees were equally unconvinced. Project managers noted that the dashboards did not address day-to-day frustrations. Analysts complained that predictive models were misaligned with client engagement. Some began calling it "the shiny toy," a sign that the tool was seen as a distraction rather than a solution.

The board pressed harder. "How does this tie to Meridiem's positioning?" one director asked. Leadership pointed to platform capabilities but struggled to connect them to outcomes that mattered in the market. The technology was advanced, but without integration, it created noise, not traction.

Then the talent problem emerged. High-potential recruits began declining offers, citing confusion about Meridiem's technology vision. "The system looks impressive," one candidate said, "but it is not clear what role technology plays in your future." At the same time, experienced IT staff started leaving for competitors where technology was embedded at the core of strategy. "It is hard to innovate here," one departing engineer remarked, "because no one agrees how technology fits."

The talent drain exposed the deeper issue. Meridiem had not reframed technology as a core variable of advantage. It was still treated as an add-on, a department-led initiative detached from leadership and strategy. Without integration, even sophisticated tools failed to create stickiness with clients, coherence for employees, or credibility in the market.

Competitors, meanwhile, pressed their advantage. Their technology was not always more advanced, but it was embedded in strategies that emphasized speed, transparency, or cost. Clients valued the coherence. Talent saw the opportunity to work in organizations where leadership, strategy, and technology reinforced one another. Meridiem, by contrast, was left with impressive features that nobody believed in.

Meridiem's story at this stage is not about the failure of a tool. The technology itself worked. The failure was architectural: Technology had been pursued in isolation—without catalytic leadership or a coherent blueprint. Instead of enabling, it distracted. The consequences included client skepticism, employee disengagement, board frustration, and a growing talent exodus. Together, they revealed the same lesson the company had yet to learn. Technology becomes the enabler of modern advantage only when it is synchronized with leadership and strategy. Without that connection, even the most advanced systems create motion without momentum, leaving the organization modern in form but static in practice.

Chapter Summary

Technology has moved from support act to central enabler. It no longer simply accelerates advantage; in many cases, it determines whether advantage can be achieved at all. When embedded in leadership direction and strategic coherence, technology turns clarity into traction. When left isolated, it consumes resources without creating distinction.

Excellence in technology is now as critical as excellence in leadership or strategy. Together, they form a synchronized system where energy, coherence, and pace reinforce one another. Differentiated strategies collapse without it, and even the most credible leaders lose relevance if technology fails to deliver. What matters is not novelty or spend but discipline. Technology embedded in leadership intent and strategic design becomes the force that converts motion into momentum.

This is why we refer to *integrated technology* as the third variable of *modern advantage*. Leadership catalyzes the equation, strategy provides the blueprint, and technology enables them both. Separated, they drift. Synchronized, they sustain coherence in motion. The next section, *Part III: Excellence Within and Synchronization Across the Variables*, opens with *Chapter 7: Excellence Within: Renewing Capability Before Alignment*, examining how organizations strengthen each variable from the inside out, building the depth of capability required before pursuing synchronization across leadership, strategy, and technology.

PART III

Excellence Within and Synchronization Across the
Variables

PART III

CHAPTER 7: EXCELLENCE WITHIN: RENEWING CAPABILITY BEFORE ALIGNMENT

Where Advantage Begins to Fade

Every leader knows the meeting that looks fine on paper. Slides are clear, numbers hold, conversation is polite. Yet the room feels slower than it should. Questions circle longer, decisions come smaller, energy drains just a bit faster than it used to. Nothing is technically wrong, yet momentum has thinned. What you are feeling in that moment is advantage beginning to fade from within.

Organizations rarely lose ground all at once. They lose it quietly, inside the disciplines that once gave them strength. Leadership starts repeating messages instead of sharpening them. Strategy drifts toward incremental comfort. Technology upgrades outpace adoption. Each function still performs, but the connection between excellence and relevance loosens. By the time the misalignment shows up across the enterprise, the erosion has already happened inside it.

This chapter begins where most recovery efforts do not: at the source. Excellence within each variable—leadership, strategy, and technology—is the foundation of modern advantage. *When that excellence dulls, no amount of cross-functional alignment can hold the system together.* The work here is less about new initiatives and more about renewal, restoring depth, credibility, and rhythm inside the core disciplines before trying to synchronize them across the organization.

Most organizations do not notice this erosion at first because it hides behind the appearance of order. Dashboards still glow green. Quarterly metrics suggest stability. Yet beneath that surface, the fundamentals that create momentum—clarity in leadership, coherence in strategy, and capability in technology—begin to lose their edge. What once felt like discipline starts to feel like drag. Meetings take longer to reach conclusions. Strategic reviews produce more updates than choices. Digital investments deliver outputs but not outcomes. None of these are crises; they are the quiet indicators that excellence within the system is slipping faster than the organization realizes.

The instinctive response is to reach for integration: to tighten coordination, build cross-functional task forces, or launch culture initiatives designed to reconnect the enterprise. Those efforts can help, but they rarely last if the foundation underneath them has weakened. Integration cannot compensate for erosion within. When the quality of leadership decisions, the distinctiveness of strategy, or the discipline of technology execution dulls, alignment across those elements can only amplify the weakness. The first step toward restoring coherence is not to add new links between the variables but to strengthen each one individually so the links have something solid to connect.

Excellence within rebuilds the foundation on which renewal depends. It is the internal rhythm that allows synchronization across the system to mean something. Organizations that ignore this step mistake visibility for vitality; they chase the appearance of integration instead of rebuilding capability at the source. The sections that follow trace that source back to its three core disciplines—leadership, strategy, and technology—to show how renewing depth within each is the prerequisite for coherence across all.

Leadership Excellence Renewed

Leadership is often the first place where excellence begins to erode, though it is rarely the first place leaders look. The decline is quiet, more rhythm than event. Decisions slow because they travel through too many checkpoints. Communication expands but clarifies less. Leaders keep saying the right things, but their words have lost their edge; they no longer create movement. What once was credibility becomes familiarity, and familiarity breeds inertia. Most organizations do not notice this shift until it has already shaped their culture. People wait for permission instead of momentum.

Renewing leadership excellence begins with capability, not charisma. Every modern leader draws from a blend of professional, personal, and technological skills. Those skills—how leaders think, communicate, decide, and adapt—form the raw material of leadership performance. What differentiates leaders who remain relevant is how those skills combine into competencies: applied fluency that allows them to operate under pressure, align across disciplines, and move with complexity instead of being slowed by it. Renewal is not about discovering new traits; it is about developing, reactivating, and recombining the skills that matter most for the conditions they now face.

The forces shaping leadership today—speed, transparency, constant visibility, and converging expectations from stakeholders—demand that those competencies evolve faster than they once did. Leaders who stay credible do not rely on experience as insulation; they treat it as context for new learning. They update their skill combinations continuously. Communication fuses with translation to turn strategy into traction. Decision-making merges with adaptability to sustain confidence amid ambiguity. Technology awareness joins with curiosity to keep judgment current. Each of these combinations reflects the same principle: Updated skills create new competencies, and new competencies restore leadership credibility.

Renewal also means shortening the distance between awareness and action. In a world that moves faster than deliberation, credibility is no longer earned by being right eventually; it is earned by being clear early. Modern leaders do not trade precision for speed, but they understand that precision means little if it arrives too late. The goal is not more decisions but better cadence: a predictable rhythm for how direction is set, tested, and reinforced. When people know how and when decisions will be made, alignment follows without the overhead of constant communication.

Finally, renewed leadership excellence shows up as coherence at the top. If the senior team is not modeling how leadership, strategy, and technology stay in rhythm, no one else can sustain it. Coherence does not require unanimity; it requires visible consistency on priorities and trade-offs. When leadership operates with that unity of rhythm, trust rebuilds quickly. People stop waiting for instruction because they can feel direction. Leadership excellence, at its core, is less about authority than about how skill becomes competence and competence becomes coherence, the chain that turns leadership from position into motion. Just as renewed leadership rebuilds credibility, strategic renewal restores direction, the blueprint that keeps that credibility pointed somewhere useful.

Strategic Excellence Renewed

Strategy loses strength differently than leadership does. It does not slow; it dulls. The edges that once defined distinction begin to round off, replaced by familiar goals and recycled language. Teams start to equate progress with activity: more projects, more analysis, more metrics, while the hard work of making real trade-offs fades into the background. Strategy becomes a process to manage rather than a direction to believe in. You can tell it is happening when planning sessions feel busy but no longer decisive, when objectives expand rather than narrow, and when everyone can describe what the organization does but few can explain what makes it different.

Modern strategy depends on three qualities: distinctiveness, coherence, and adaptability. None of them hold on their own. They require renewal. Renewing strategic excellence begins with clarity of intent. Advantage today comes not from being everywhere but from being deliberate about where not to be. Renewal requires revisiting the organization's design for distinctiveness, the combination of choices, capabilities, and customer experiences that make it difficult to imitate. The most effective leaders treat this exercise as a test of coherence rather than ambition. They simplify until the strategy reveals what truly drives value and what no longer does. They prune initiatives that compete for the same resources or dilute focus, knowing that coherence—not volume—creates traction.

Modern strategy also depends on adaptability, the ability to stay consistent in purpose while flexible in execution. Markets now move too quickly for static plans to hold, and advantage decays faster than planning cycles can adjust. Strategic renewal, therefore, means building a discipline of refreshment, short and frequent recalibrations that keep direction relevant without rewriting the entire plan. It is less about annual off-sites and more about continuous translation, turning new information into course corrections that preserve intent while adapting its expression. Strategy remains the blueprint, but it is now a living one, revised in motion rather than in retreat.

Excellence in strategy also requires reconnecting it to the real work of the organization. The farther strategy drifts from where value is created, the faster it loses meaning. Renewal brings strategy back to the front lines by making it visible in decisions, investments, and trade-offs. When people can see how their work reflects strategic intent, alignment stops being a slogan and becomes a habit. When that link breaks, even well-designed strategies collapse into noise. Renewal closes that gap, ensuring that what leadership articulates the enterprise can actually deliver.

Strategic excellence is ultimately the discipline of staying decisive in complexity. It is less about forecasting the future than about creating clarity in the present, making choices that reinforce one another, remaining adaptable as conditions evolve, and sustaining coherence between what is promised and what is practiced. When strategy regains that rhythm, leadership has something credible to activate and technology has something precise to accelerate. Renewal within strategy does not make organizations faster; it makes them surer, anchored in intent but agile in execution. If strategy defines the path, technology determines the pace. Renewal within technology ensures that clarity does not outrun capability.

Technology Excellence Renewed

Technology loses its edge in an even quieter way than either leadership or strategy. It rarely fails outright; it fades into the background, still functional but no longer formative. Systems keep running, but they no longer differentiate. Dashboards multiply, but the information they surface no longer guides better decisions. Technology becomes something the business maintains rather than something it uses to move. The clearest sign of erosion is when leaders start describing technology as stable. Stability is valuable for infrastructure; however, it is fatal for advantage.

Renewing technology excellence begins with restoring fluency, not investment. Modern advantage depends less on what tools an organization owns and more on how well its leaders understand, question, and apply them. The strongest organizations cultivate technology fluency, an enterprise-wide grasp of how technology enables pace, scale, and learning. That fluency turns awareness into capability. When leadership teams talk comfortably about data models, automation, or AI ethics, technology stops being a siloed language. It becomes the shared medium through which strategy and execution connect.

The foundation for that fluency was established earlier. Technology is the enabler of modern advantage, providing pace, precision, and connection when integrated with leadership and strategy. As with the other variables, excellence inside technology does not sustain itself; it must be renewed. Renewal starts by embedding technology deeper into the flow of work. Instead of adding new platforms or rebranding old ones, organizations reinforce capability by placing technology at the point of value creation, where customers experience the brand, where data improves judgment, and where systems learn from outcomes. Renewal happens when technology evolves in the same rhythm as business design, not on a separate calendar.

Another dimension of renewal is maintaining the balance between modernization and mastery. Many organizations chase modernization as proof of progress, new systems, new releases, new spend, but forget that mastery is what turns modernization into advantage. Renewal requires both, upgrading infrastructure while ensuring people can actually use it. The measure of excellence is not the number of tools deployed; it is the speed and confidence with which teams use them to make better decisions. When technology improvement outpaces adoption, investment turns into inertia. Renewal restores proportion, aligning innovation with comprehension so the organization remains fluent, not overwhelmed.

Technology excellence also depends on rhythm. The leaders who sustain it treat modernization as an ongoing cadence rather than an episodic event. They build structures that synchronize updates, capability building, and feedback loops across functions. They do not wait for transformation programs; they make progress habitual. Over time, that rhythm becomes cultural. Teams expect evolution, not replacement. Technology becomes less about hardware and more about habit, the consistent discipline of integrating new capability into the organization's daily motion.

When renewed in this way, technology excellence reclaims its role as the enabler of modern advantage. It accelerates the clarity that leadership provides and amplifies the coherence that strategy defines. When renewal lags, technology becomes noise, an impressive array of tools searching for purpose. When renewal holds, it becomes the connective tissue that keeps the entire system alive, ensuring that the organization moves as one at the speed the market now requires.

When Renewal Creates Lift

Earlier in the book, we warned against chasing symptoms instead of causes. Renewal within the variables is what it looks like to treat the cause, and the first sign that the treatment is working is lift. As leadership, strategy, and technology recover depth, the organization begins to move again, even before formal alignment takes hold. That movement feels subtle at first: Decisions get made faster, meetings end with more conviction, and projects that once lingered start to close. Nothing has been reorganized, but something has shifted. The system is breathing again.

Lift shows up differently in each discipline. In leadership, it is the return of confidence, the sense that decisions carry weight again. In strategy, it feels like focus, the noise of competing priorities begins to fade. In technology, it is ease, processes that once dragged start to glide. None of these shifts are dramatic, but together they signal that the organization is regaining traction. It is the early stage of rhythm, when excellence within begins to sound like harmony.

That lift comes from small consistencies that appear naturally when excellence within the variables starts to compound. A clearer leadership cadence sharpens how priorities are communicated. Renewed strategy simplifies where resources go. Technology fluency removes friction from delivery. Each improvement may seem minor in isolation, but together they reduce noise, rebuild trust, and restore a sense of shared motion. These connections are not planned; they happen because the quality of the work itself has risen. When depth returns, alignment stops feeling forced, and rhythm starts to reappear.

This is the moment when organizations begin to feel coherence before they can measure it. The structure is the same, but the tone has changed. People no longer talk about alignment; they practice it in how they respond, how they decide, and how they hand work to one another. The lift is not the result of new systems; it is the natural consequence of competence reconnecting across boundaries. Those early renewals create the foundation for synchronization later. By the time formal alignment efforts begin, the culture already carries momentum. Excellence within has made the organization lighter, faster, and ready to move as one.

Meridiem: When Depth Erodes

Several months after Meridiem's widely publicized strategy reset, the company appeared to be moving again. New digital pilots were underway. Leadership had launched a training program for emerging managers. The technology function was reporting faster sprint cycles, and internal communications showcased early wins. On paper, everything looked productive. Yet in board updates and client meetings, the results told a different story. Revenue was flat, margins continued to narrow, and project close rates were slipping. Progress existed, but it was not adding up to momentum.

Inside the company, the work felt heavier, not lighter. Each function had acted on the directive to improve excellence within its own domain—but in isolation. The leadership team focused on decision discipline, introducing new governance templates and meeting cadences. Strategy refreshed its priorities, adding a sharper focus on customer retention. Technology invested in analytics and automation, promising greater visibility across projects. Each effort was credible in its own right, yet none of them connected. Improvements made locally were canceling each other out globally.

That fragmentation showed most clearly in the company's largest account. During a renewal presentation, a senior executive promised the client faster delivery enabled by new analytics capabilities. The technology team, learning of this only afterward, clarified that the platform was not ready for deployment. Operations, eager to protect the relationship, offered discounts to preserve goodwill. The client accepted the concession but not the confusion. What was meant to demonstrate progress instead revealed drift. The account remained, but the client's confidence did not.

By the next quarter, similar patterns appeared elsewhere. Teams were working longer hours, yet deadlines slipped. Managers spoke about alignment, but their plans contradicted one another. Strategy documents and system dashboards told parallel stories that never met. People were working harder, not smarter, and they knew it. As one manager admitted quietly, "We are busy in every direction, but none of it feels forward."

The organization had rebuilt effort, not excellence. Its metrics suggested movement, but its reality was maintenance. What had once been a confident rhythm had become a hum of disconnected activity. Meridiem's leaders could sense the fatigue but could not yet name its cause. They were repairing the pieces without strengthening the whole. Until the depth inside leadership, strategy, and technology was renewed, progress would remain cosmetic, something the company talked about more than it experienced.

In her next board update, the CEO wrote only one line: We are busy, but we are not better. It was the first clear acknowledgment that Meridiem's problem was not effort but evolution. The company had been applying its old strengths to new realities, and only by modernizing leadership, strategy, and technology could it rebuild the capability its environment now required.

Chapter Summary

Advantage rarely disappears in a single moment. It erodes quietly inside the disciplines that once created it. This chapter examined how that erosion takes shape and how renewal restores the foundation for modern competitive advantage. Leadership, strategy, and technology each lose strength in different ways: credibility gives way to familiarity, distinctiveness fades into activity, and modernization drifts into maintenance. Renewal within each re-establishes capability, coherence, and rhythm. Excellence within is what keeps advantage renewable; it turns the pursuit of improvement into the precondition for alignment.

Organizations that catch these internal drifts early avoid larger fractures later. Paying attention to the small, local warnings—such as delayed decisions, diluted strategy, or tools that outpace adoption— allows leaders to correct before complexity compounds. Renewal begins not with new structures but with sharper fundamentals: updated skills that form modern competencies, disciplined choices that sustain distinction, and technology fluency that keeps pace with strategy. When those elements strengthen together, the system naturally starts to regain lift even before formal synchronization begins.

The next challenge is to ensure that renewed depth translates into shared motion. Integration across the variables only holds when each is healthy on its own, but coherence across them is what converts renewal into momentum. Chapter 8, *Synchronization Across: Turning Depth into Rhythm,* explores how that happens, how alignment becomes an active discipline, and how rhythm—not structure—transforms individual renewal into collective traction.

CHAPTER 8: SYNCHRONIZATION ACROSS: TURNING DEPTH INTO RHYTHM

When Motion Becomes Rhythm

Renewal creates lift, but lift without focus quickly disperses. Synchronization gives that lift purpose: It turns movement into rhythm—and activity into advantage. When leadership, strategy, and technology begin to operate in time with one another, priorities sharpen. The energy that once scattered across initiatives starts to collect around what truly matters. Decisions align not because people are told to agree, but because the logic behind those decisions has become visible and shared.

Synchronization is more than coordination. Coordination rearranges schedules; synchronization establishes cadence. It connects the three variables to a common tempo so that leadership's direction, strategy's design, and technology's execution reinforce each other instead of competing for attention. The result is not more motion but smarter motion, effort guided by agreement on what deserves momentum and what can wait. In synchronized organizations, prioritization becomes the clearest expression of alignment. It is where coherence is felt, not just discussed.

As rhythm takes hold, something subtler happens: Confidence returns. People begin to trust the flow of decisions again. Even when trade-offs are hard or outcomes uncertain, they can sense that choices are connected to a single direction. The anxiety that comes from competing priorities gives way to a shared steadiness. Synchronization replaces the noise of individual agendas with the reassurance of collective intent. It tells everyone, in effect, that we are moving in the same direction for the same reasons.

This confidence is what separates activity from advantage. Busy organizations move constantly but rarely in unison; synchronized ones move deliberately and compound their effort through timing. Rhythm converts renewal into reliability, the sense that progress is not accidental but repeatable. It is the discipline that keeps clarity from fracturing as conditions change and ensures that the organization's energy remains concentrated on the work that moves it forward.

Confidence in rhythm also changes how accountability feels. In unsynchronized organizations, accountability often shows up as control: leaders checking, confirming, and correcting. In synchronized ones, accountability becomes shared awareness. People hold themselves to the rhythm because they believe in it. They see how their own timing affects others and protect that flow instinctively. Confidence does not reduce discipline; it makes discipline collective. That is the cultural dividend of synchronization: The system polices itself through pride, not pressure.

The Discipline of Synchronization

Synchronization is not a program or a quarterly exercise; it is a continuous behavior that keeps the organization moving to a single rhythm. When leadership, strategy, and technology move together, they create an internal tempo that makes priorities self-evident. Leaders do not have to over-explain or micromanage because people can see the pattern of decisions and understand where effort belongs.

Synchronization does not slow the system down; it establishes predictable pace. It is what allows speed to remain coordinated rather than chaotic.

At its core, synchronization rests on three operating disciplines: cadence, clarity, and connection.

Cadence is the timing mechanism, the steady pulse that turns planning into progress. Regular reviews, transparent check-ins, and short feedback loops create continuity without ceremony. Cadence keeps the system honest; when the clock is predictable, accountability follows naturally.

Clarity is the communication mechanism, the shared understanding of why and how decisions are made. It removes interpretation as a variable. Clear priorities are repeated often enough to become instinct so teams can adjust without waiting for new directives.

Connection is the feedback mechanism, the loop that carries information between leadership, strategy, and technology. It turns static coordination into dynamic learning, allowing one variable to update the others before misalignment spreads. This rhythm also refines how leaders allocate capital. Synchronization surfaces where investment will compound and where it will cancel. When the three variables stay in conversation, budgets start to mirror reality in motion. Resources flow toward momentum, not noise. Projects that once competed for funding are now sequenced rather than stacked. Leaders can say no with confidence because everyone understands the logic behind the sequence. Synchronization does not just organize work; it organizes money, attention, and time around the same pulse.

Together, these disciplines filter ambition through reality. Every organization has more ideas than it can fund or execute. Synchronization turns that abundance into order. It converts intent into sequence: what happens first, what follows, what waits. It is the filter that ensures investment flows to initiatives where all three variables intersect: credible leadership sponsorship, coherent strategic fit, and enabling technology. Effort outside that intersection is not wasteful; it is simply deferred until it contributes to the rhythm.

Prioritization becomes intelligent when it reflects synchronization rather than proximity or politics. In synchronized organizations, resources concentrate around traction instead of noise. Leaders can see which initiatives are accelerating the system and which are draining it. This shared visibility gives confidence that trade-offs are being made for the right reasons. Synchronization does not eliminate conflict; it channels it. It provides a structure where debate clarifies rather than fragments, and where decisions hold because they are understood in context.

Rhythm, once established, becomes a form of trust. Teams learn to anticipate when direction will update, when feedback will loop back, and when priorities will shift. That predictability frees them to act decisively between cycles rather than waiting for confirmation. Synchronization is not a signal to slow down; it is the reason everyone can speed up safely. It turns alignment from agreement into motion—and motion into momentum.

Operating in Concert

Synchronization comes to life when the three variables—leadership, strategy, and technology—begin to act less like independent functions and more like instruments in the same ensemble. Each has a different sound, but the performance depends on how they play together. Leadership establishes tempo, strategy provides the score, and technology supplies the amplification that carries the sound through the organization. When the rhythm holds, you can feel it. Meetings shorten. Trade-offs get made faster. Projects hand off cleanly instead of stalling in transition. The work feels lighter because it is moving in time.

Employees often describe this shift before leaders do. They talk about finally knowing what matters, or that meetings end earlier but decide more. It is not sentimentality; it is sensory. The rhythm changes the atmosphere. The organization feels quieter yet faster, as if background noise has dropped. That sensation is more than morale; it is evidence that synchronization is working at the human level. When people can anticipate the rhythm, they contribute more fully to the performance.

Operating in concert begins with how decisions are sequenced. Leadership sets the tempo by defining the pace of review and renewal. Strategy follows by aligning the sequence of initiatives to that tempo: what to advance now, what to incubate, what to stop. Technology then calibrates its efforts to match, embedding tools and data where the rhythm is already strongest rather than chasing novelty. The outcome is intelligent focus: Resources flow naturally toward initiatives that already demonstrate momentum. Energy collects around what works instead of dispersing across what is merely available.

When synchronization takes hold, prioritization becomes effortless because it is visible in the work itself. Teams no longer compete for attention or resources; they can see which efforts amplify the shared rhythm and which are out of tune. This shared visibility builds confidence across the enterprise. People understand not only what is being prioritized but why. That understanding transforms compliance into conviction. Teams stop asking whether their project matters; they already know it does, or they know why it can wait.

Feedback completes the loop. As synchronized organizations operate, technology captures real-time data that feeds strategy, and strategy translates those insights into decisions leadership can act on. The cadence of review turns into a heartbeat: input, decision, execution, learning, then back to input again. Each pass through the loop sharpens focus and reinforces trust. Over time, this rhythm compounds advantage. The organization no longer fights for alignment; it operates in coherence, making decisions with confidence because they are made with common direction in mind.

The practical test of synchronization is simple: when change arrives, does the organization adjust or pause? Synchronized enterprises adapt mid-stride without losing tempo. They do not wait for stability to resume; they create it through rhythm. That ability to move together, to pivot, prioritize, and progress without losing confidence, is what distinguishes busy organizations from those that are truly smart.

Barriers to Rhythm

Even capable organizations struggle to maintain rhythm once the tempo increases. Synchronization can be fragile; it depends on shared timing, mutual attention, and a measure of humility, the willingness to yield pace when another part of the system must lead. Those qualities fade quickly under pressure. The most common barrier is not disagreement about goals but the quiet re-emergence of silos. Leadership begins to sprint ahead with new ambitions before strategy can reframe direction. Strategy starts producing analysis faster than technology can adapt the systems that make those ideas real. Technology continues to modernize, but without clear strategic priority, improvement turns into drift. The rhythm breaks not because the music stops but because each part begins to play in its own key.

When rhythm fades, prioritization splinters. Each function defends its own version of what matters most. Leadership funds initiatives that signal progress; strategy refines models that promise foresight; technology pursues upgrades that demonstrate innovation. None of these are wrong, but without synchronization, they compete for attention and capital. The result is activity without traction, resources allocated to motion rather than momentum. Synchronization prevents this slide by forcing trade-offs that are collective, not local. It ensures that every dollar, every hour, and every decision contributes to the same underlying tempo.

Another barrier lies in the illusion of clarity. Leaders often mistake alignment for synchronization; they achieve agreement on priorities but fail to establish shared timing. The strategy is right, the projects are sound, but everything happens out of sequence. Decisions arrive after budgets are locked. Technology capacity lags behind strategic ambition. Communication updates faster than execution. These timing gaps erode confidence. People start second-guessing decisions not because they disagree but because they no longer trust the rhythm that produced them. Synchronization solves this by anchoring clarity in cadence; when timing is predictable, confidence returns.

Rhythm also falters when feedback slows. Synchronization thrives on real-time learning; it weakens when data, insight, or response time lengthen. The organization begins to manage the past instead of adapting to the present. Fixing this does not require new dashboards; it requires restoring the feedback loops that keep the variables in conversation. Leadership must listen as actively as it directs. Strategy must treat execution as evidence, not aftermath. Technology must translate information into insight, not just output. When these feedback cycles accelerate, rhythm resets.

Recognizing these barriers matters because synchronization rarely fails loudly. It decays quietly in the spaces between variables, where timing slips and trust fades. The solution is not more process; it is more presence. Leaders who preserve rhythm do so by staying alert to those small shifts in tempo: by realigning cadence before the music drifts too far to hear.

The earliest signs of renewed synchronization are easy to overlook because they are felt more than measured. Decisions start arriving in sequence. Hand-offs shorten. Conflicting priorities disappear without escalation. Technology projects close on time because their purpose stays stable. Most telling, conversation changes; people reference timing as often as tasks. They speak in terms of next, after, and wait until this lands. These are the linguistic fingerprints of rhythm taking hold. When timing language replaces urgency language, synchronization has begun to embed itself.

Meridiem: When the Pieces Start to Align

Two quarters after Meridiem's leadership retreat, the company felt different. The work was still intense, but the noise had softened. Weekly meetings that once overflowed with competing priorities now followed a clear sequence: leadership cadence first, strategy adjustments second, technology deployment third. For the first time, the order made sense. People could see how one variable depended on the others and how moving in time produced better outcomes than sprinting alone.

The first visible proof came from a stalled modernization initiative. The technology group had been building a client analytics platform for months without clear alignment on how it would drive growth. Instead of pushing harder, they paused to synchronize. Leadership restated the commercial goal to improve client retention. Strategy clarified which customer segments mattered most. Only then did technology adjust its roadmap to support those targets. The next release launched on schedule and, more importantly, solved the right problem. The lesson was quiet but unmistakable: speed had finally begun to serve strategy, not undermine it.

Confidence spread faster than the technology. Managers began to trust decisions they had not personally influenced because the reasoning behind them was visible. Strategy reviews no longer required rehearsed defenses; the cadence itself created alignment. Even the finance team noticed the change, with fewer last-minute requests and fewer overlapping budgets. Synchronization had replaced persuasion. The conversation shifted from "who owns this" to "where does this fit." That single change in language signaled that Meridiem's leaders were beginning to think as one.

The real turning point came when the leadership team agreed on what to stop. Three overlapping pilot programs were frozen in a single meeting, not as cost cuts but as sequencing choices. Resources were redirected to the areas already showing traction. The relief across the organization was palpable. People no longer equated progress with motion; they began to understand it as rhythm. Momentum, once forced, was starting to feel natural.

Meridiem was not fixed, but it was showing signs of synchronization across the organization. Decisions carried a logic that people could follow, and that made them believe again. The organization had learned that coherence does not come from consensus; it comes from cadence. Each variable still had more work to do, but the rhythm was real. For the first time in a long while, Meridiem was moving with confidence, in time and in the same direction.

Chapter Summary

Synchronization transforms renewal into rhythm. It turns the depth rebuilt within leadership, strategy, and technology into coordinated motion that compounds instead of competes. Through cadence, clarity, and connection, synchronization links individual improvement to collective progress. When timing aligns, priorities sharpen. When priorities sharpen, confidence returns. The organization begins to act with one pulse: moving faster, deciding cleaner, and investing where energy multiplies rather than divides.

The discipline of synchronization filters ambition through shared intent. It orders what happens first, what follows, and what waits. This sequencing is what distinguishes activity from advantage. Busy organizations move constantly but without coherence; synchronized ones move deliberately, concentrating effort where the three variables intersect. That focus restores confidence that decisions, even the hard ones, are being made with a common direction in mind. Synchronization does not remove uncertainty; it replaces it with trust in the rhythm that guides the enterprise forward. For leaders, synchronization becomes the quiet architecture of credibility. It shows up less in speeches than in the consistency with which priorities, funding, and feedback arrive on time.

The next step is to recognize how synchronization shows up in practice: how an organization can tell when the rhythm holds and when it is beginning to drift. The next chapter, *Reading the Signals: Seeing Coherence in Motion,* explores what synchronization looks like from the outside in—the cultural tone, the performance patterns, and the early signs that reveal whether rhythm is compounding or eroding.

CHAPTER 9:
READING THE SIGNALS:
SEEING COHERENCE IN
MOTION

From Rhythm to Signals

Synchronization does not stay hidden for long. Once rhythm takes hold inside an organization, it begins to show itself through signals that make coherence visible in motion. These signals appear in decisions that land with consistency, in messages that sound familiar no matter who delivers them, and in outcomes that arrive on time without drama. What began as alignment among leadership, strategy, and technology becomes visible in how the organization behaves. Rhythm, when sustained, produces signals, evidence that coherence has moved from aspiration to habit.

Every organization emits these signals, whether it intends to or not. They are not the same as localized cues that reveal issues inside a single variable; signals operate between disciplines, the by-products of timing, tone, and follow-through. Some signals confirm that the system is synchronized: meetings that end with clear decisions instead of deferred ones, projects that close on schedule because dependencies were understood, customer experiences that feel consistent even as offerings evolve. Other signals warn that rhythm is starting to drift: priorities multiplying, communication growing louder but less clear, decisions taking longer to hold. The art of leadership at this stage is learning to read both.

Reading signals is how leaders sense coherence in motion. It is less about dashboards and more about pattern recognition, seeing the small consistencies that compound into trust and the small delays that turn into drag. Healthy organizations develop signal literacy, a shared ability to interpret what the rhythm is telling them. They can diagnose drift before data confirms it, and they know when the system needs tuning rather than overhaul. In this way, signals become both mirror and instrument panel, the reflection of how well synchronization is working and the guidance for keeping it alive.

The value of signal awareness is not only diagnostic; it is directional. Signals tell leaders where rhythm is strongest and where it is at risk. They highlight where to invest, where to pause, and where the system itself can self-correct. In synchronized organizations, this awareness becomes cultural. People notice when timing feels off, when decisions no longer arrive in sequence, when a tone begins to change. They do not wait for leadership to announce misalignment; they surface it. When rhythm becomes signals, the organization has reached a higher form of maturity, coherence that is not managed but sensed.

Internal Signals: Culture and Cadence

The first signals of coherence appear inside—not outside—the organization. Before customers notice consistency or investors reward stability, people inside can feel when timing starts to align. Earlier in this book, we described cues, localized warnings that something inside a single variable might be weakening. Signals are different. They are macro outcomes, not micro alerts. Cues live inside disciplines; signals live between them. They do not point to an isolated issue but reveal the overall health of synchronization. Culture is the earliest broadcast of that health. It does not declare itself with slogans or metrics; it shows up in tone, cadence, and shared confidence. When rhythm is healthy, culture feels calm but focused, candid without being combative. People spend less energy interpreting intent and more energy moving in time with one another.

The clearest internal signal is cadence, the tempo of interaction and decision. In synchronized organizations, meetings begin and end on time because the rhythm is predictable. Information moves without friction because the intervals between planning and action are consistent. When cadence holds, the organization develops a collective sense of time. People know when feedback will arrive, when decisions will be revisited, and when priorities will rotate. Predictability does not make work mechanical; it makes it sustainable. Teams can plan with confidence because they trust the beat will hold.

Language provides another reliable signal. In synchronized cultures, the vocabulary of urgency gives way to the vocabulary of rhythm. People speak in terms of next, after, and ready, not immediately, yesterday, or ASAP. The change seems small, but it signals maturity. Urgency implies panic; rhythm implies flow. You can tell a great deal about an organization's health by how it talks about time. When conversations shift from how fast to how sure, coherence is strengthening.

Behavioral signals complete the picture. In synchronized environments, people prepare for meetings differently because they trust the process. They do not over-document to defend a position; they arrive ready to advance the decision. Leaders listen more than they summarize because information has already circulated through the system. Silence in meetings no longer means disengagement; it means alignment. The culture runs quieter because fewer things need re-litigation. This quiet is not complacency; it is coordination. It is what happens when everyone knows the next beat.

The most powerful signal of all is psychological safety in motion. When rhythm holds, people speak up earlier, not louder.

They trust that surfacing a concern will not be misread as dissent but as calibration. The rhythm itself creates that safety, the predictability of when and how issues will be addressed. Teams do not fear being wrong because they know the next chance to adjust is coming soon. Coherence replaces defensiveness with contribution. That shift is subtle, but it is how rhythm translates into culture, through consistency that becomes confidence and confidence that becomes trust.

External Signals: Performance and Perception

Synchronization eventually becomes impossible to hide. What begins as internal rhythm soon radiates outward through the organization's interactions with customers, partners, and markets. External audiences cannot see cadence directly, but they can feel its effect in reliability, responsiveness, and tone. Just as culture broadcasts the health of rhythm inside, performance and perception broadcast it outside. They are the visible outcome of timing held and coherence sustained.

The first external signal is reliability, the consistency with which the organization delivers what it promises. Reliable companies do not need to advertise their dependability; customers and partners experience it. They receive updates when expected, products when scheduled, and answers that align with prior commitments. Reliability becomes a strategic asset not because it guarantees perfection but because it reduces friction. In markets where speed and noise dominate, reliability stands out as confidence made tangible. Stakeholders trust synchronized organizations because the rhythm of delivery conveys control without rigidity.

The second signal is responsiveness. In synchronized organizations, the ability to react quickly does not come from adrenaline; it comes from rhythm. Teams adapt to new conditions because decision pathways are already clear. Technology surfaces information in time, strategy provides context, and leadership converts it into motion. When disruption hits, synchronized organizations adjust mid-stride without losing tempo. From the outside, this looks like agility. On the inside, it is simply cadence doing its job. Customers notice when companies stay composed under pressure. That composure is rhythm performing in real time.

Communication provides another outward signal. Coherent organizations speak with one voice even across many channels. Messages to customers, partners, and investors sound consistent not because they are scripted but because they originate from shared understanding. In synchronized systems, communication travels at the same tempo as action. What leaders say publicly mirrors what employees hear internally. The alignment of words and behavior becomes its own form of credibility, a rhythm the market can hear. Incoherence, by contrast, sounds like noise, mismatched statements, shifting narratives, or promises out of sync with delivery. When communication drifts from cadence, trust erodes.

Performance data itself becomes a signal, though not in the conventional sense. Metrics in synchronized organizations tell a story of balance. Growth is steady rather than spiked, profitability improves without sacrificing innovation, and customer satisfaction rises in parallel with employee engagement. These patterns do not occur by accident; they are statistical evidence of rhythm. Just as a heartbeat can be read on a monitor, organizational coherence leaves a measurable trace. Leaders who study their own performance patterns can detect synchronization by the absence of volatility. Predictability, once dismissed as dull, is often the strongest proof of momentum that lasts.

The final and most telling signal is perception: the way the organization is described by those it affects. Customers call synchronized companies easy to work with. Partners describe them as steady or dependable. Investors see them as disciplined rather than conservative. These impressions form organically, through repetition over time. Perception becomes a lagging indicator of rhythm's maturity; when outsiders describe the company in ways that match how insiders experience it, coherence has become complete.

External signals matter because they confirm that rhythm has moved beyond management into identity. They show that the system's timing is no longer enforced but embodied. When coherence can be felt from the outside, the organization is not just synchronized, it is trusted. That trust converts rhythm into reputation and transforms performance into proof of advantage.

Interpreting Drift

Rhythm rarely breaks all at once. It slips. A small delay in one area becomes hesitation in another, and before long, the music feels off even though everyone is still playing the same notes. The early signs of drift are almost imperceptible from a distance, but inside the organization, they can be sensed long before they can be seen. Learning to interpret drift is what allows leaders to tune the system before it falters, to correct timing rather than rebuild tempo.

The first indicator of drift is a distortion in timing—when decisions arrive either too early or too late for the rhythm they are meant to serve. An initiative that leaps ahead of strategy, or a technology update that outruns adoption, both create dissonance. The sequence may still look correct on paper, but the intervals between actions have stretched. In synchronized organizations this usually starts with good intent: someone trying to move faster, but even small timing mismatches accumulate. When the rhythm that once guided pacing begins to fragment, coordination starts to feel like friction.

A second indicator is a rise in communication noise. As rhythm loosens, communication gets louder. Meetings multiply, emails lengthen, and updates become more frequent but less useful. What feels like diligence is often compensation for lost trust in timing. People fill the gaps with words because the system's cadence no longer feels reliable. Leaders may notice that decisions need more reminders or that progress reports include more context than content. These are cultural echoes of drift, the organization speaking over its own rhythm to stay in sync.

The third signal is re-litigation: the reappearance of decisions once thought settled. In synchronized systems, clarity and cadence make direction self-reinforcing; when they fade, old debates resurface. Teams start revisiting priorities, questioning trade-offs, or reopening budget discussions without new information. This is not defiance; it is the absence of conviction that the previous decision will hold. Re-litigation wastes time, but more importantly, it erodes confidence in leadership's tempo. When people can no longer predict which choices are permanent, they hesitate to move at all.

Energy imbalance is the fourth and often final indicator. You can feel it in meetings where some functions accelerate while others stall, or when enthusiasm for new projects outweighs attention to existing commitments. Synchronization depends on shared energy across the variables; when one outpaces the rest, drag appears elsewhere. Energy imbalance is how most organizations experience drift first, not as failure but as fatigue. People start working harder to achieve the same effect. The system is still performing, but it is now powered by effort instead of rhythm.

The ability to interpret drift is the difference between maintenance and mastery. *Healthy organizations do not fear these signals; they rely on them.* Drift does not mean synchronization has failed; it means it is doing its job by revealing where the rhythm needs adjustment. Leaders who listen for distortions in timing, rising noise, re-litigation, or imbalance can intervene with precision rather than disruption. They retune the system, not reinvent it. In this way, synchronization becomes self-correcting, a living discipline that learns from its own variations to stay in time.

Meridiem: Reading the Signals

Six months after Meridiem began synchronizing its leadership, strategy, and technology rhythms, the company's progress had become visible. It did not arrive through major announcements or breakthroughs but through tone. Conversations were quieter, meetings shorter, and updates more predictable. People were beginning to sense rhythm rather than chase it. Projects advanced in sequence, and decisions landed with less debate. The organization was not faster in the traditional sense; it was steadier. That steadiness felt like momentum.

The leadership team noticed the difference first in their reviews. The volume of discussion had decreased, but the quality had risen. Reports arrived before they were requested. Metrics aligned naturally with strategy not because of new dashboards but because the cadence of decision-making had normalized. It was subtle, fewer surprises, fewer apologies, fewer urgent flags. Leaders could tell that coherence was holding, even if they could not yet quantify it. Synchronization was producing signals they could finally read.

Not every signal was positive, and that was the point. The team began to recognize the early signs of drift, slower follow-up from certain regions, inconsistent client communication in one business line, and a rise in quick-sync meetings that had once disappeared. None of these were crises, but each was a faint distortion in timing. Instead of launching a new initiative, the executive team used their standing cadence to address it. Strategy leads recalibrated priorities, and technology shifted resources toward better workflow visibility. Within weeks, the noise receded. Meridiem had tuned the rhythm without breaking it.

The realization that mattered most was not about performance; it was about perception. Managers noticed that employees had started describing the company differently. "It just feels easier to get things done," one said. "I know who is deciding what—and when." Those remarks were not part of a survey or program; they were the cultural echo of synchronization. People could feel decisions arriving in sequence, and that gave them confidence that direction was holding.

Meridiem had learned to read its own signals. Some were steady, others uneven, but all were visible. The organization no longer mistook silence for drift or activity for progress. It could tell instinctively when coherence was holding and when it was beginning to stretch. Rhythm had become self-reinforcing, visible and understood, the sign that Meridiem had finally learned to listen to itself.

Chapter Summary

Signals are rhythm made visible. They reveal how coherence behaves once synchronization becomes habit, through tone, timing, and the consistency of outcomes. This chapter explored how those signals appear both inside and outside the organization. Internally, culture and cadence broadcast rhythm through language, predictability, and trust. Externally, reliability, responsiveness, and perception translate that rhythm into credibility. Together, these signals form the proof that coherence is not only achieved but sustained.

Reading signals is the discipline that keeps synchronization alive. Organizations that can interpret their own patterns, such as distortions in timing, rising noise, re-litigation, or energy imbalance, can correct drift before it becomes dysfunction. Signal literacy replaces guesswork with awareness. Leaders learn to read rhythm like data, understanding that small variations carry meaning and that coherence requires continuous tuning. When organizations listen to what their rhythm is telling them, they gain the confidence to adapt without losing timing.

The next stage is to ensure that synchronization and signal literacy do not depend on individual vigilance but on institutional design. In the next chapter, *Institutionalizing the Approach: Making Advantage Durable*, examines how organizations embed these disciplines into governance, communication, and renewal systems so that performance becomes self-sustaining and advantage endures through continuity, not constant intervention.

CHAPTER 10: INSTITUTIONALIZING THE APPROACH: MAKING ADVANTAGE DURABLE

From Practice to Foundation

Modern competitive advantage is not built from a single burst of performance. It comes from an approach that can be practiced, refined, and repeated. The organizations that sustain distinction over time are the ones that turn this approach into a foundation, something that lives underneath strategy and leadership transitions, not beside them. The question is no longer how to build advantage once but how to build it again and again without starting over.

Every organization eventually faces that test. Early successes come from energy, focus, and leadership attention. Over time, those qualities fade or shift to new priorities. What remains determines whether the approach endures. When modern competitive advantage is institutionalized, renewal, synchronization, and adaptation stop being projects; they become the normal way of working. The company does not rely on extraordinary effort to regain coherence; it expects coherence to be there.

Building that foundation requires simplicity, not structure. The goal is not to add layers of process but to establish a small number of habits that anchor the approach. The first is a shared understanding of purpose, the answer to why advantage exists and what makes it distinct. The second is rhythm, the discipline of staying in step as decisions move from leadership to strategy to execution. The third is reflection, the practice of looking back just long enough to carry learning forward. When these three habits hold, the system renews itself naturally. It remains steady through change because modern competitive advantage has been absorbed into how people think, decide, and lead.

Building that kind of foundation does not happen through one transformation project; it develops through small consistencies. Leaders model the behavior first, how they frame choices, how they connect decisions, how they pause long enough to explain the why behind the work. People learn the rhythm by watching it performed. The approach spreads not through rollout but through repetition. Once that pattern takes hold, the foundation becomes self-strengthening; every decision reinforces the way the next one will be made.

Institutionalizing the approach is therefore an act of subtraction, not addition. It removes dependency on individual style, temporary focus, or external urgency. It leaves behind a foundation that makes progress predictable without making it rigid. When an organization can apply modern competitive advantage repeatedly, through new leaders, new markets, and new cycles, it proves that advantage is not a moment in time. It is a way of operating that lasts.

Building the Foundation for Repeatability

An approach becomes repeatable when the work that sustains it stops depending on personality or perfect conditions. The strongest organizations make that shift by focusing on a few essentials: knowing what matters most, making decisions in a consistent way, and keeping attention on the things that give the company its edge. These are not complex ideas, but they are difficult to protect once success creates noise and distraction. The foundation that holds them in place is built through focus, sequence, and follow-through.

Focus means protecting what the organization does best. Over time, priorities expand and messages multiply. The company that once stood for something clear begins to sound like everyone else. The foundation for repeatability begins when leadership restores focus to the few things that define advantage and lets the rest fall away. This is not about narrowing ambition; it is about giving the organization a center of gravity strong enough to withstand change.

Sequence keeps work in order. In companies that practice modern competitive advantage well, direction, design, and delivery still happen quickly, but they happen in the right order. Leadership sets intent, strategy shapes how that intent becomes distinct, and technology brings it to life. When that sequence holds, progress compounds instead of colliding. When it breaks, even good ideas lose momentum. Repeatability depends on keeping that order visible and intact.

Follow-through turns decisions into habits. Many organizations know what to do; fewer stay with it long enough for the approach to take hold. Follow-through is what converts rhythm into reliability. It is the quiet discipline of checking that choices are carried through, not just announced. Over time, that steadiness becomes the company's signature. Customers begin to expect it. Employees begin to rely on it. Advantage stops being something the company tries to create and becomes something it naturally renews.

The foundation for repeatability does not require new systems or slogans. It requires attention to the basics that made the organization distinct in the first place and the resolve to protect them as it grows. When focus, sequence, and follow-through are preserved, the approach to modern competitive advantage no longer needs to be relaunched. It simply continues. Leaders who protect that simplicity turn out to be the quiet architects of repeatability. They do not write playbooks; they guard intent. Their role is to make sure the work that built advantage never becomes invisible. By talking about choices, not slogans, they keep the approach practical and remind people that clarity, not intensity, is what makes progress hold.

Sustaining Balance Across the Variables

Modern competitive advantage depends on balance. The three forces that create it—leadership, strategy, and technology—must continue to move together. When one begins to outrun the others, rhythm breaks and repeatability fades. The challenge is not deciding which variable matters most; it is keeping them in proportion as the organization grows, new leaders emerge, and markets shift. The weight will always move, but the balance can hold if leaders know what to watch.

The first imbalance usually appears when leadership overextends. Strong leaders bring clarity, energy, and speed, but they can also create dependency. The approach becomes person-shaped rather than principle-shaped. When the leader moves on, progress pauses. Organizations that sustain advantage keep leadership distributed. They share the responsibility for direction so rhythm does not hinge on one voice. Clear roles, consistent forums, and transparent decisions make leadership portable. The company keeps its bearings even when faces change.

The second risk comes from strategy becoming too abstract or too reactive. Strategy provides shape and distinction; it connects purpose to the marketplace. When it drifts toward planning for its own sake, or shifts too frequently to match headlines, it loses its anchoring role in the equation. Sustainable organizations keep strategy practical and visible. They test choices in the work itself. They use real results to confirm whether distinctiveness still holds. Strategy remains steady not because it ignores change, but because it translates change into choices everyone can understand.

Technology can fall out of balance in two ways: by racing ahead or by lagging behind. In both cases, it becomes detached from the rhythm that makes it valuable. The organizations that sustain advantage use technology to serve direction, not define it. They invest where technology amplifies leadership intent and sharpens strategic distinction. They pause where it distracts or fragments. In balanced organizations, technology decisions are timing decisions; they are made to keep the whole approach in motion, not to impress the market with novelty.

Maintaining this balance requires perspective more than process. Leaders who keep the variables aligned do not chase symmetry; they look for harmony. They sense when one element is pulling too hard or moving too slowly and adjust before it becomes visible as drift. Over time, that steadiness becomes the company's signature sound, leadership, strategy, and technology playing in time, none louder than the others. That balance is what makes the approach repeatable; it ensures that advantage remains integrated even as the world around it changes.

The steadiness that balance creates also frees leaders to look outward again. Once the variables move together reliably, attention can shift back to customers and markets, the reason the approach exists in the first place. Internal balance becomes the condition that allows external distinction to keep advancing.

Renewal as a Discipline

Every advantage fades unless it is refreshed. The mistake most organizations make is treating renewal as an event—a response to pressure or a change in leadership—rather than as part of the normal rhythm of the work. Renewal done in cycles looks like disruption; renewal done as discipline looks like stability. *Companies that sustain advantage build the habit of small, steady corrections instead of waiting for crisis to force a reset.*

Sustained renewal begins with curiosity. People at every level are encouraged to ask whether what once created distinction still does. The question is not a challenge to direction; it is how direction stays alive. When leaders make that reflection routine, through reviews, conversations, and observation, the organization keeps learning without slowing down. The work of improvement becomes invisible because it is constant.

The discipline of renewal also depends on humility. No approach, no matter how strong, stays current on its own. Markets shift, customers evolve, and technology changes pace. Organizations that hold their rhythm through all of that do not assume permanence; they assume movement. They keep structures light enough to bend, decisions open enough to update, and people close enough to reality to see when the world has moved first. Humility makes rhythm sustainable because it allows adjustment without loss of confidence.

Finally, renewal holds when it feels ordinary. When people know that the approach to modern competitive advantage will always be reviewed, re-tuned, and reinforced, change loses its drama. The organization stops waiting for permission to adapt and starts expecting to. That expectation is what makes advantage repeatable. It proves that the work of leadership, strategy, and technology is never finished, but it is always familiar, a continuing process of keeping what is distinct in step with what is next.

Meridiem: When Advantage Becomes Habit

A year after Meridiem's leadership retreat, no one talked about frameworks or transformation anymore. The language of programs had disappeared. The company was not introducing a model; it was living one. The rhythm that once required attention had become instinct. Meetings still began with the same review cadence. Strategy updates still closed with three questions. Is this distinctive? Is it repeatable? Does it connect? Technology plans still followed in sequence, translating ideas into capability. What had started as a deliberate effort now felt like muscle memory.

The difference was most visible in small things. Managers no longer waited for alignment meetings to make trade-offs; they already knew the sequence of decision rights. Teams stopped escalating issues that belonged in their own lane; they solved them in rhythm and moved on. Product teams adjusted roadmaps within weeks of new market signals because data, strategy, and leadership were already connected. Progress looked uneventful but felt steady, which was exactly the point. The company had traded drama for dependability. For Meridiem's leaders, the greatest relief was seeing stability come from habit rather than oversight. They no longer pushed the rhythm forward; they simply protected it, trusting the organization to carry the beat on its own.

Customers noticed the change before Meridiem did. Feedback described projects that "just worked," communication that "felt clear," and solutions that "arrived when they were needed." None of those comments mentioned process or cadence; they described confidence. Meridiem's rhythm had reached the outside world. The organization was not faster than its competitors; it was simply more consistent. That consistency had become its advantage.

Inside the company, leaders used the same calm language to describe performance: steady, clear, predictable. They no longer held special sessions to re-teach the approach; new employees learned it by experience. Decisions arrived in order, momentum stayed even, and corrections happened quietly in stride. What had begun as the pursuit of alignment had turned into the practice of continuity.

Meridiem had not perfected the pursuit of modern competitive advantage; it had normalized it. The approach no longer depended on heroics, workshops, or declarations. It was sustained by habits that felt ordinary but produced results that were not. Advantage had become routine, renewed and reinforced in every cycle of work. The company had learned the final lesson of rhythm, once it becomes habit, distinction follows naturally.

Chapter Summary

Advantage becomes sustainable when the approach behind it can be practiced again and again without starting from zero. This chapter described how modern competitive advantage matures from effort into foundation, how organizations move from doing the work of alignment to living it. When rhythm becomes familiar and coherence becomes habit, modern competitive advantage stops being a project and becomes a way of operating. Renewal, synchronization, and reflection no longer require instruction; they happen naturally because the approach has taken root.

Repeatability depends less on structure than on steadiness. The organizations that preserve advantage over time protect a few simple disciplines: clear intent, orderly decision-making, and consistent follow-through. They keep leadership, strategy, and technology in balance even as markets shift. Renewal is expected, not exceptional. Change feels like continuity because the same approach that created distinction continues to renew it. The company no longer relies on energy or personality to stay in motion; the approach itself supplies the rhythm.

The next step is to apply that rhythm in context, to see how modern competitive advantage adapts to different industries, scales, and ambitions, and in doing so, converts into a distinct advantage. *Part IV: When Modern Competitive Advantage Becomes a Distinct Advantage* with *Chapter 11, The Inherent Value in Pursuing Competitive Advantage,* explores how leaders take the same approach that made advantage sustainable and make it personal. They shape it to their environment so that what endures inside the organization becomes unique in the world it serves.

PART IV

When Modern Competitive Advantage Becomes a
Distinct Advantage

PART III

CHAPTER 11:
THE INHERENT VALUE IN PURSUING COMPETITIVE ADVANTAGE

Why Pursue Advantage

Up to this point, this book has focused on what modern competitive advantage is and how it works. We have examined how leadership, strategy, and technology must evolve and move together to create a repeatable rhythm. Those ideas have shaped the foundation of the modern approach.

Now, in practice, that same approach takes on a new dimension. *When it is applied within your unique environment, it becomes your distinct advantage,* the same foundation, owned and expressed in your context. The pursuit that once felt conceptual now becomes personal.

The reason to pursue this advantage is simple. Applying the modern approach within your organization's realities creates options. It opens doors that remain closed to everyone else. It gives organizations access to better opportunities, stronger partnerships, and more profitable growth. It turns market pressure into leverage. When an organization practices the approach with consistency, it earns the ability to attract customers, talent, and investment on its own terms. It drives revenue, builds margins, and protects performance when conditions turn. Over time, that rhythm of practice turns universal principles into distinctive capability, what this book calls a distinct advantage.

Without an intentional approach, progress becomes fragile. Organizations compete on price instead of value, chase opportunity instead of shaping it, and work harder for smaller returns. They rely on effort rather than influence. Applying the modern approach in context changes that balance. It lets capability compound instead of decay. Over time, that consistency becomes identity. The approach strengthens reputation, signals credibility and confidence, and makes it easier for customers to choose, employees to stay, and investors to believe.

The nature of advantage itself has evolved. Traditional advantage was often built through control of scale, assets, or distribution. Modern advantage depends on adaptability, clarity, and connection. It rewards those who can learn, adjust, and deliver in rhythm with their environment. The fundamentals of ambition remain the same, but the path to achieving them has changed. The modern approach provides the mechanism through which these universal dynamics take shape in context and become distinct advantage in practice.

This is why the pursuit matters. Not every company will dominate its industry, but every company benefits from trying. The pursuit of modern competitive advantage strengthens focus, builds adaptability, and creates momentum. It improves decision quality and shortens the distance between ambition and results. Even when market leadership remains out of reach, the pursuit itself builds the foundation that, once applied inside your organization, becomes a distinct advantage of your own.

We are having this conversation here, at this point in the book, because this is where advantage becomes organizational. It is also where the modern approach begins to convert into a distinct advantage, when it stops being a universal framework and starts becoming the way you operate. At this level, advantage is not abstract; it is intensely practical. It is about your organization winning business, keeping customers, growing markets, and sustaining confidence.

When practiced deliberately, the approach that produces a distinct advantage remains the most reliable currency in business. It earns attention, secures loyalty, and buys time to adapt. It creates the space to lead rather than follow. The next question is what happens when that pursuit never starts, or when it no longer qualifies as a priority.

When Advantage Isn't Pursued

Most organizations do not lose advantage in a single moment. They simply stop practicing a repeatable approach that creates it. What begins as a pause in focus becomes a pattern of maintenance. Over time, energy that once created traction is redirected toward protecting what already exists. Momentum fades, not through failure but through comfort.

When the approach to building an advantage is not pursued, progress becomes incremental. Strategy begins to sound familiar because it is built from what the organization already knows. Decisions take longer. Meetings revolve around optimizing what is rather than creating what could be. The organization still performs, but the edge that once distinguished it begins to blur. The absence of practice weakens rhythm, and without rhythm, the approach cannot evolve into a distinct advantage.

Customers notice first. They stop describing the company as different and start describing it as reliable. Reliability is valuable, but in a volatile market it is rarely enough. Talent notices next. Ambitious people look for environments that stretch them. When a company stops pursuing the approach, it signals that the learning curve has flattened. Those who want to grow will eventually find somewhere else to do it.

At the financial level, the absence of pursuit shows up quietly. Margins narrow as competitors replicate old strengths. Sales teams lean harder on discounts. Marketing begins to defend perception rather than extend reach. The numbers can still look solid, but they start requiring more explanation. Each quarter takes slightly more effort to deliver than the one before.

Culturally, organizations that stop practicing the approach often confuse activity with motion. Projects multiply, yet direction feels unclear. Leaders work harder to create energy, but the energy no longer compounds. It dissipates. The organization becomes busy maintaining itself.

Some companies never begin the pursuit at all. They operate effectively but always in response to others. They benchmark, follow, and improve on what is visible. Their progress depends on someone else's lead. Without an internal drive to apply the modern approach, they inherit the tempo of the market rather than setting it.

The effect is not collapse; it is erosion. The organization remains standing but smaller in influence, less distinct in presence, and slower to adapt. What was once initiative becomes inertia. The failure to practice the approach means the organization never reaches the point where advantage can become distinct. It loses not only momentum but also the opportunity for its strengths to take shape in a way that is uniquely its own.

The pursuit of modern advantage was never meant to be a periodic overhaul. When the approach is practiced within your environment, it becomes a continuous discipline that shapes identity. Failing to pursue it is not only the loss of rhythm; it is the loss of distinction.

The Approach Needed to Pursue Modern Advantage Has Evolved

For much of the last century, organizations built advantage through control. They gained strength by owning resources, defining markets, and defending position. Scale and efficiency were the primary levers of success, and leadership measured progress through stability and reach. That approach worked when environments were predictable and when time rewarded those who could hold their ground.

Those conditions no longer exist. Complexity now moves faster than control. Technology, transparency, and stakeholder expectations interact continuously, reshaping competition in real time. What once created stability now creates drag. *Advantage can no longer be managed as a fixed position; it must be pursued as a continuous process that learns and adjusts in motion.*

This evolution is what preserves the value of pursuit itself. The way advantage is built has changed, but its importance has not. The same evolution that reshaped advantage also made distinction possible. When organizations learn to apply the modern approach within their own structure, culture, and pace, they stop imitating and start differentiating. Adaptability becomes identity. That is the point where modern competitive advantage, once universal in design, becomes a *distinct advantage* through practice in context.

The modern approach that builds coherence depends on two connected disciplines, excellence within and synchronization across leadership, strategy, and technology. Each must first achieve strength on its own. These are not the traditional definitions of the three disciplines but their modern expressions, leadership that creates clarity through adaptability, strategy that defines direction through coherence and distinction, and technology that connects and accelerates both. Excellence within each builds the credibility and depth that give an organization confidence and resilience.

Depth alone, however, no longer holds. The forces shaping markets move together, and the disciplines that create advantage must do the same. Synchronization across leadership, strategy, and technology converts internal excellence into motion. It keeps the organization coherent as it adapts and ensures that progress compounds rather than collides. Without synchronization, even strong parts compete for attention. Without excellence, synchronization has nothing solid to connect. Advantage now depends on the balance between these two conditions, depth inside and rhythm across.

This evolution changes how organizations pursue progress. In earlier eras, renewal arrived through large and infrequent efforts such as new strategies, restructures, or technology transformations. Today, those cycles are too slow. Modern pursuit favors smaller and deliberate renewals that keep excellence current and rhythm intact. The work is constant but not heavy. Once the rhythm is established, it becomes self-reinforcing. Strength within each discipline supports alignment across them, and alignment deepens that strength in return.

Leadership's role is to preserve coherence across the rhythm. It sets the pace of renewal and ensures that modernization in one area does not outrun the others. Strategy provides direction for that renewal, translating ambition into clear choices that can be executed and defended. Technology connects and amplifies both, giving the organization the reach and precision that modern markets demand. Together, they form an operating rhythm built on credibility, connection, and pace.

This is why the pursuit of modern advantage matters. Its value is inherent; sustained pursuit builds capability, focus, and confidence that endure beyond the results themselves. The work has evolved from episodic reinvention to continuous renewal. What once required extraordinary effort can now be maintained through discipline. The method has changed, but the purpose remains the same, to create distinction that endures because it adapts.

This is the evolution of pursuit. The next section explores how the modern approach functions in practice and how, when it is applied within your environment, it becomes your organization's *distinct advantage*.

When the Modern Approach Becomes Distinct

The modern marketplace rewards organizations that can stay coherent while the world around them accelerates. Pace and complexity now define competition, and speed alone no longer differentiates performance. What separates the strongest organizations is their ability to maintain excellence within leadership, strategy, and technology while keeping those disciplines synchronized across the enterprise. The result is coherence that holds under pressure, direction that adapts as conditions change, and capability that connects ambition with execution.

This is the essence of the modern approach. It reflects how advantage now behaves: dynamic, continuous, and sustained through rhythm rather than control. In this environment, advantage is no longer a trophy earned once and defended for years. It is a process built on two linked disciplines, depth inside each of the core variables and rhythm across them. Excellence within creates credibility and strength. Synchronization across converts that depth into movement and momentum. Together, they form the cycle that keeps advantage renewable.

When that cycle is practiced consistently within one organization's context, its culture, its decisions, and its customers, it becomes more than a system of renewal. It becomes the organization's unique expression of the modern approach, a rhythm others can see but cannot easily replicate. That is the point where the modern approach, applied and owned in context, becomes a distinct advantage.

Organizations that embrace this pursuit discover that progress becomes steadier even as their environment becomes more volatile. Leadership sets the pace and coherence of renewal. Strategy channels that coherence into decisions that feel relevant and forward-looking. Technology amplifies both, translating clarity into delivery at the speed markets now expect. Each discipline preserves its depth, yet together they function as one. That integration, more than any single breakthrough, defines the modern pursuit of advantage and is the foundation for distinction in practice.

This way of working also changes how work feels inside the organization. Teams spend less time chasing alignment because alignment is embedded in how decisions are made. Meetings shorten because priorities are already understood. Initiatives connect more easily because technology carries information and intent through the same flow. Progress depends less on intensity and more on rhythm. The organization no longer runs on urgency; it runs on cadence.

Externally, this shows up as credibility and stability. Customers experience consistency even as offerings evolve. Partners and investors sense confidence because results arrive predictably. Employees recognize an enterprise that moves quickly without losing clarity. What the market perceives as brand strength is coherence made visible, the cumulative proof that leadership, strategy, and technology are working together in motion.

The value of pursuing modern advantage through this approach lies in its sustainability. It delivers renewal without disruption and strength without exhaustion. Once the rhythm of modernization and alignment takes hold, maintaining it becomes part of daily leadership. Progress continues, but it no longer feels extraordinary. The enterprise becomes lighter, more deliberate, and more resilient.

The pursuit of advantage remains vital because the approach behind it has evolved into a discipline that any organization can apply. It no longer requires dramatic transformation to stay relevant; it requires consistency, attention, and integration practiced continuously but proportionately. When applied within a unique environment, the modern approach evolves into a distinct advantage. It turns the pursuit of competitive advantage from an episodic effort into an ongoing capability that keeps organizations coherent in motion and distinct in the marketplace.

Chapter Summary

Competitive advantage continues to matter because it remains the most reliable measure of organizational health. What has changed is both the nature of the pursuit and the expression of the outcome. The equation of modern competitive advantage still anchors the work, but when the modern approach is practiced within your unique environment, it evolves into a distinct advantage, your own rhythm of coherence and renewal. It stops being a framework to reference and becomes a way of operating.

This is the point where method becomes identity. The same disciplines of leadership, strategy, and technology now work through your cadence, your decisions, and your context. What was once the pursuit of alignment becomes the practice of character. The organization's coherence turns into its signature, the visible expression of how the modern approach lives within its world.

From this point forward, the conversation has shifted from definition to application. *Chapter 12: An Approach That Fits Every Environment* explores how the modern approach adapts across contexts, showing how the same foundation that creates coherence in one organization becomes distinct in another through practice in its own environment.

CHAPTER 12:
AN APPROACH THAT FITS
EVERY ENVIRONMENT

Built for Difference

Every organization is different. This is both the most obvious truth in business and the one most often ignored when new ideas appear.

Some operate in markets that change daily, while others move through long cycles shaped by regulation or capital intensity. Some are led by founders with clear vision, others by boards that balance complex stakeholder agendas. Some compete for customers, others for funding, others for trust. No two contexts are identical, and no single structure or playbook fits them all.

Leaders know this instinctively. It is the first objection that surfaces whenever a new idea appears. That may work there, but it would never work here. The concern is understandable. The differences that define an organization—including its history, industry, governance, and appetite for risk—shape every decision it makes. They determine what is possible, how fast it can move, and what trade-offs it must manage. To ignore those differences is to misunderstand how organizations actually work.

Acknowledging that reality is the starting point. The intent of this chapter is not to minimize uniqueness but to recognize it fully, and then show that the modern approach to the pursuit of advantage was built for it. The approach does not flatten difference; it depends on it. Its strength lies in how it adapts across environments without changing its logic. It works in global corporations and regional enterprises, in mission-driven organizations and market disruptors, because it focuses on the interactions that exist inside every organization, not the surface details that make them distinct. When practiced in context, the same approach that creates modern advantage becomes the foundation for a distinct advantage unique to each environment.

The modern approach rests on three disciplines: leadership, strategy, and technology. These disciplines exist in every organization regardless of industry or scale. The expression of each may vary, but the need for coherence among them does not. Leadership still defines credibility and pace. Strategy still determines focus and coherence. Technology still connects capability to action. These fundamentals hold true whether the work is performed in hospitals, factories, professional services firms, or public agencies.

What changes from one environment to another is weighting and rhythm. In heavily regulated sectors, strategy often anchors the system, guiding leadership and technology within tighter boundaries. In fast-growth settings, leadership carries more weight, setting tempo and absorbing the turbulence that comes with expansion. In technology-led industries, the platform itself dictates pace, requiring strategy and leadership to synchronize around constant release cycles. The proportions shift, but the process remains the same. Excellence within each discipline and synchronization across them keep the organization coherent. That is why the approach travels so well, and why it becomes distinct when practiced in context.

This is not theory; it is the lived pattern of organizations that continue to thrive while others lose pace. Their environments differ, but the conditions that define competition—including speed, transparency, complexity, and expectation—are shared. The pursuit of modern advantage works because it organizes these shared conditions into a rhythm that any organization can sustain. It does not ask leaders to ignore their realities. It gives them a way to navigate those realities with greater coherence and less friction.

When the approach is practiced consistently within one organization's unique environment, it becomes more than a system of renewal. *It becomes that organization's distinct advantage, a rhythm that others can see but cannot easily replicate.* The truth that every organization is different is not a challenge to the approach; it is its reason for being. Practiced in context, the approach creates clarity within complexity, translates difference into structure, and ensures that variation in environment does not become variation in performance. Understanding this is the first step toward seeing how the same process, applied within your environment, produces results that are both consistent and distinct.

The Forces Every Organization Faces

While every organization operates in its own context, all are shaped by the same underlying forces. These forces are not selective. They cut across industries, geographies, and business models. They define the environment in which the modern approach to advantage must operate. Understanding them reveals why the same foundation can succeed everywhere, even as its expression varies.

The first shared force is pace. Markets now move continuously rather than periodically. Information flows instantly, technology evolves daily, and customers adjust expectations without warning. The time available to sense, decide, and act has collapsed. Whether the organization operates in finance, manufacturing, education, or government, the same truth applies. Advantage decays in the space between recognition and response. The organizations that endure are those that move with rhythm rather than haste and that renew themselves before pressure forces them to.

The second shared force is complexity. Competition no longer arrives from one direction at a time. Economic cycles, regulatory shifts, social expectations, and technological disruptions overlap and amplify one another. Leaders cannot isolate problems as neatly as they once could. The modern environment behaves as a network in which every choice interacts with others. Success now depends less on controlling variables and more on connecting them coherently. This is why alignment across leadership, strategy, and technology has become the central requirement of performance.

The third shared force is transparency. Stakeholders, employees, investors, and communities now see deeper into organizations than ever before. Every decision is visible, and every inconsistency is amplified. Reputations rise or fall in real time. In this context, credibility has become as measurable as profitability. Leadership must maintain authenticity and clarity, strategy must reinforce purpose with evidence, and technology must make information accessible and trustworthy. Together, they determine how believable an organization's story sounds once exposed to the world.

The fourth shared force is expectation. Every stakeholder now demands more—and faster. Customers expect personalization and immediacy. Employees expect purpose and adaptability. Investors expect both growth and governance. Communities expect responsibility as well as performance. These expectations differ in form but share one essence. They evolve faster than the structures built to meet them. Organizations that respond only when expectations peak will always appear behind. Those that adapt continuously stay credible even when they are still learning.

These four forces—pace, complexity, transparency, and expectation—form the common environment that no organization can escape. They are the reason the modern approach to advantage works in every setting. While every enterprise interprets them through its own realities, the underlying conditions are the same. Leadership, strategy, and technology remain the instruments through which these forces are managed and turned into coherence. The organizations that master this balance perform better not because they face fewer challenges, but because they have learned how to stay aligned as those challenges evolve.

Recognizing these shared forces closes the distance between difference and universality. They prove that the world every organization inhabits may appear unique, but its pressures are shared. The modern approach provides the structure to manage those pressures in rhythm. When practiced inside a specific environment, it turns that shared rhythm into a unique expression of strength. That is how modern competitive advantage becomes a distinct advantage, one shaped by universal forces but defined by context.

Why a Modern Approach Works in Any Environment

The modern approach to advantage works in every organization because it is built on fundamentals that do not change when the environment does. It succeeds by strengthening the disciplines that define how all organizations function and by keeping those disciplines connected. It does not ask leaders to conform to a single model. It gives them a structure that flexes to fit their own realities.

Every enterprise, regardless of sector or scale, draws coherence from the same three disciplines: leadership, strategy, and technology. Their relative weight may shift, but their relationship does not. Together, they form the structure that turns awareness into performance.

A start-up in renewable energy may rely on leadership to set vision and attract capital, while strategy and technology evolve in motion. A large bank may rely on strategy to anchor decisions within regulation, with leadership and technology adjusting around it. A professional services firm may depend on technology to drive speed and insight, requiring leadership and strategy to align more quickly. In each case, the order and emphasis shift, but the logic remains. Excellence within each discipline creates credibility, and synchronization across them creates traction.

This is why the approach fits everywhere. It is not a formula that prescribes what to do; it is a structure that guides how to work. It does not replace the uniqueness of a business model or the character of a culture. It strengthens both by aligning them with the forces that shape every market. The organization does not lose its individuality by adopting the approach; it becomes clearer about what makes that individuality valuable.

The modern approach adapts to context because it is built for motion. It recognizes that leadership, strategy, and technology will never move at exactly the same speed, but they can still move in rhythm. It accepts that conditions will keep changing, yet coherence can be preserved. It treats alignment as discipline rather than as a project and turns renewal into a normal part of the operating rhythm.

When the approach is practiced effectively, difference becomes an advantage rather than an excuse. Distinct markets, cultures, and risk profiles create new ways to express the same principles. The method scales because it is based on connection rather than conformity. It works as well in a single business unit as it does across a multinational enterprise. What changes is not the model but the way leaders use it to interpret their world.

The strength of the modern approach lies in its universality without uniformity. It holds steady in any setting because it is built on fundamentals that never disappear, the need for credible leadership, coherent strategy, and enabling technology. When this approach is applied within a unique environment, it does more than sustain coherence; it transforms it into distinction. That is the moment when modern competitive advantage becomes a distinct advantage, shaped by universal principles but defined by context. Each generation of leaders rediscovers these truths under new conditions, and each time they do, they find that difference is not an obstacle to advantage. It is the context that gives advantage meaning.

Where the Work Begins

Every organization begins the pursuit of modern advantage from a different point. Some start with leadership because clarity and credibility must come first. Others start with strategy because direction is in question or differentiation has faded. A growing number begin with technology because it offers the fastest route to scale or connection. None of these starting points are wrong. What matters is that the work begins, and that the other disciplines are brought into motion quickly.

The pursuit of advantage through the modern approach is not a linear process. It does not require perfect readiness in every dimension before progress can be made. Organizations that wait for ideal conditions rarely move. The strength of the approach is that it allows pursuit to begin wherever leverage exists. It can flex to context without losing coherence. Leadership, strategy, and technology are interdependent, so progress in one discipline naturally pulls the others forward.

Starting with leadership often creates immediate traction. Clarity from the top shortens decision time, aligns teams, and reveals where strategy and technology are out of rhythm. Starting with strategy builds discipline around focus and trade-offs. It brings order to activity and defines what will create relevance. Starting with technology builds reach and pace. It exposes friction, improves transparency, and forces clarity about process and decision rights.

Every organization has a natural place to begin. A regulated enterprise may start with strategy to meet new conditions. A founder-led company may start with leadership to prepare for scale. A global manufacturer may start with technology to connect dispersed operations. Each path is valid if the work expands to include all three disciplines. The pursuit of modern advantage succeeds not because of where it starts but because of how quickly it becomes whole.

Progress also depends on momentum. Small wins build confidence, and confidence attracts energy. Early movement in one discipline gives the organization something to point to, a visible sign that change is underway. Those visible signs matter. They prove that advantage is not theoretical but built in real time. Momentum turns intent into belief, and belief into habit.

Starting where you are is not about convenience; it is about pragmatism. It recognizes that every organization faces different constraints, yet all share the same need for coherence. The question is not whether to pursue advantage but how to begin. The work does not have to start perfectly. It simply has to start, and once it does, momentum will carry the organization forward as the disciplines strengthen one another in motion.

When the approach is practiced this way—beginning wherever leverage exists and expanding until leadership, strategy, and technology move in rhythm—it becomes more than a process. It becomes the organization's distinct advantage, the natural outcome of applying the modern approach within its own context and pace. Through that practice, coherence turns into identity, and motion becomes the organization's signature rhythm.

Chapter Summary

Every organization is different. The variables that shape performance—including market, regulation, scale, culture, and history—create unique environments and expectations. Those differences matter. They determine how fast an organization can move and what trade-offs it must make. Yet they do not exempt any organization from the pursuit of advantage. The modern approach to advantage exists precisely because of this variation. It provides a structure that adapts to difference without losing coherence, allowing organizations to work within their realities rather than against them. When applied in context, that same structure evolves into a distinct advantage, the organization's own rhythm of coherence and renewal.

While environments differ, the forces shaping competition are the same. Pace, complexity, transparency, and expectation define the conditions under which all organizations now operate. The modern approach fits them all because it strengthens the disciplines that every enterprise depends on, credible leadership, coherent strategy, and enabling technology. These disciplines remain universal even as their expressions vary. When excellence within each meets synchronization across them, difference becomes a source of strength rather than limitation.

The pursuit of modern advantage succeeds when it begins where each organization has leverage and expands from there. What matters is not the starting point but the rhythm that connects leadership, strategy, and technology as the work progresses. When the modern approach is practiced inside an organization's unique environment, it becomes that organization's distinct advantage, the visible expression of universality made personal. The next chapter, *Chapter 13: Starting Where You Are,* illustrates how this process unfolds across different contexts, showing that while every path is unique, the outcome is the same, organizations that move in rhythm and sustain advantage that is both modern in method and distinct in form.

CHAPTER 13: STARTING WHERE YOU ARE

When the Work Takes Shape

Every organization begins or renews the pursuit of modern advantage from a different place. Some start because something has broken, others because something has stalled, and a few because they can see an opportunity before it fully arrives. The conditions vary, but the intent is the same, to build strength that lasts longer than a single market cycle.

This chapter brings that idea to life. It shows what it means for organizations to start or restart where they are, to begin the pursuit of advantage from the conditions, constraints, and opportunities that already define their world. Each example represents a different entry point into the same process. Together, they demonstrate that progress is less about where the work begins and more about how the disciplines connect once motion starts.

The modern approach makes that possible because it works from wherever an organization happens to be. It does not depend on a perfect sequence or an ideal set of conditions. It depends on motion. Progress begins with whichever discipline has the most leverage, leadership, strategy, or technology, and then expands as the other two are pulled forward. The order is less important than the connection that follows.

This is the point where theory meets reality. In practice, advantage is rarely built in a straight line. The pursuit begins or renews where urgency and readiness overlap, and from there, the work of coherence begins. The stories in this chapter show what that looks like in different environments. Each begins or renews the pursuit from a unique starting point but follows the same pattern, progress in one discipline creates traction in the others until rhythm takes hold.

Some organizations start with leadership because focus and confidence are the foundation for any change. Others begin with strategy because direction has become diffuse or coherence has faded. Many now begin with technology because capability is where customers and competitors feel advantage first. A growing number take on the evolution of all three at once, knowing that the pace of change no longer allows for sequence.

Each of these paths is valid. Each one illustrates that the pursuit of modern advantage is not about starting in the same place but about moving in the same way. The examples that follow are not blueprints to copy. They are proof that the same process adapts to any environment when leaders commit to connection across leadership, strategy, and technology as the work progresses. The previous chapter showed that the modern approach applies everywhere. This one shows what that truth looks like in motion and how, when practiced in context, it becomes a distinct advantage.

Leadership-Led Beginnings

For many organizations, the pursuit of modern advantage begins or renews through leadership. Clarity from the top is often the first sign that the organization is ready to move again. It is where belief, credibility, and pace take shape. But belief alone is not enough. Renewal starts when leaders insist on developing the modern skills that combine into competencies capable of sustaining rhythm under pressure.

Leadership-led beginnings rarely occur because leadership was weak. They emerge when the environment advances faster than leadership capability. Executives who once relied on judgment and experience discover that the skills that built success now merely preserve it. They recognize that leadership itself must evolve if the organization is to move with modern conditions. The question is not whether leaders are committed but whether they are equipped.

A global logistics enterprise faced this realization when automation began reshaping its operations. The CEO refused to treat the issue as structural. Instead, she launched a leadership development agenda centered on decision-making, curiosity, and problem-solving, skills that together form the competency of adaptability. Executives practiced real-time decision cycles using live data, learning to adjust judgment as variables shifted. Within six months, planning windows shortened, coordination improved, and delivery reliability increased. The gain came not from new systems but from new skill integration that turned speed into traction.

A global energy company faced a similar challenge. Its growth was steady, but its reputation for innovation had faded. The CEO concluded that the gap was not in technology or strategy but in leadership credibility. Communication had become cautious and resilience had weakened under public and regulatory scrutiny. A development initiative brought senior leaders together to combine communication, authenticity, and resilience—the elements that produce credibility when conditions tighten. Through open dialogue and peer coaching, executives learned to face external pressure without retreating into process. Within a year, cross-functional collaboration improved, and the organization regained the confidence to lead in emerging markets. Credibility had been rebuilt as a collective competency.

A regional financial institution confronted a subtler challenge. Growth was steady, but energy had faded. The CEO asked her team to identify which skills connected empathy to execution. The exercise revealed gaps in motivation, emotional intelligence, and coaching—skills that together create alignment. A leadership learning cycle paired executives with branch teams to practice these behaviors in daily operations. The result was sharper focus, faster feedback loops, and renewed morale. The organization rediscovered momentum by translating empathy into motion.

Leadership-led beginnings rarely create advantage on their own, but they start the motion that makes it possible. Renewal begins with capability, the deliberate development and integration of modern skills into competencies such as adaptability, credibility, and alignment. Once leaders expand what they know and how they combine it, strategy gains precision, technology finds traction, and coherence returns as the rhythm of progress.

Regardless of whether leadership is new or experienced, confident or uncertain, the same truth applies. Every organization, even those that believe they have always been pursuing advantage, must evolve through the modern approach. Leadership remains the first variable of motion, but coherence is what sustains it. Advantage strengthens only when leadership, strategy, and technology move together, continuously adapting to the world around them.

In other organizations, the first sign of movement appears through strategy, as clarity of direction becomes the next expression of readiness.

Strategy-Led Beginnings

For some organizations, the pursuit of modern advantage begins or renews through strategy. Direction becomes the catalyst for motion. When markets converge, when purpose feels blurred, or when success has bred complacency, redefining what the organization stands for and where it will compete often reawakens clarity. Strategy is the discipline that transforms scattered effort into focus and turns focus into progress toward a distinct advantage.

A global consumer goods company had diversified into too many categories. Growth looked strong on paper, but profits were shrinking and the culture had become fragmented. A new chief strategy officer initiated a reset. Instead of chasing every possible segment, the company concentrated on markets where its credibility and innovation could still create distinction. That clarity reenergized leadership, simplified decision-making, and redirected technology investment. Within a year, product teams were smaller, innovation cycles faster, and internal confidence higher. Focus replaced volume as the measure of progress, and the company began to regain the coherence that would evolve into a distinct advantage.

A fast-growing technology consultancy reached a point where expansion began to dilute identity. The firm had entered too many markets too quickly and risked losing its differentiation. The founding partners launched a strategy renewal to define a single unifying value proposition built around outcomes, not services. The new focus guided hiring, pricing, and capability development. Growth slowed briefly, then accelerated on stronger margins. Strategy restored discipline without constraining ambition. Client retention improved and referrals increased, proving that focus can expand opportunity when it sharpens relevance and coherence, the conditions that give rise to a distinct advantage.

A health system that had expanded through acquisitions faced a different challenge. Each hospital operated as an independent entity with its own leadership structure, information systems, and culture. Patients experienced the network as a collection of facilities rather than a unified organization. The leadership team began a strategy-led renewal centered on a shared vision of patient experience and care quality that transcended individual institutions. Once that intent was established, leadership alignment followed, and long-delayed technology integration became possible. Strategy gave the enterprise a single story that employees could believe in and customers could feel, renewing the rhythm that sustains advantage over time.

Strategy-led beginnings share a common thread. They start with sharper intent and end with stronger connection. Focused choices create clarity, and clarity creates speed. When organizations define what truly makes them different, they release energy that had been trapped in complexity. But clarity of direction alone is not enough. Strategy can ignite advantage, but only the modern approach can sustain it.

Whether an organization is defining its direction for the first time, renewing a strategy that has grown tired, or evolving one that has long delivered success, the same condition applies. Leadership, strategy, and technology must operate in rhythm. Every organization, regardless of maturity or confidence, must evolve through this approach. Distinction today is not achieved by having a strategy, but by how seamlessly that strategy connects to the disciplines that deliver it.

For some, capability becomes the natural place to build momentum, especially when technology offers the fastest route to renewal.

Technology-Led Beginnings

In many organizations, the pursuit of modern advantage begins or renews through technology. Capability becomes the first visible lever for change. When markets digitize, when customer expectations accelerate, or when data begins to outpace decision-making, technology often becomes the most accessible starting point for renewal. It provides tangible progress where clarity or direction may still be forming.

Technology-led beginnings occur for two reasons. In some organizations, technology has simply become too central to ignore. In others, it has advanced faster than the surrounding disciplines. The result is the same. Technology creates motion, but that motion can either generate coherence or amplify noise. The difference lies in whether leadership and strategy evolve in step with it.

A regional retailer recognized that its customers were interacting with the brand primarily through digital channels. The company had strong leadership and a well-defined strategy, but its technology had lagged behind. When the pandemic accelerated online demand, technology became the first and most urgent variable to address. The organization began by rebuilding its customer platform, investing in analytics, and automating fulfillment. The changes were visible within months, but the more significant shift happened behind the scenes. As technology capacity grew, strategy discussions changed. The leadership team began making decisions based on real-time insight rather than quarterly reports. Technology created the conditions for coherence, pulling the other disciplines forward. Online sales doubled within the first year, confirming that coherence behind the scenes quickly translates to momentum in the market.

A large university faced a different kind of pressure. Enrollment in traditional programs had declined while demand for flexible learning increased. Technology became the lever for renewal. Leadership invested in an integrated digital learning platform that connected students, faculty, and employers. The shift was technological in form but strategic in consequence. New partnerships formed, curricula evolved, and revenue streams diversified. Technology redefined the institution's relevance without changing its mission. Enrollment stabilized, and new international students joined through virtual programs, proving that technology can expand reach without diluting mission.

A manufacturing company that had long relied on engineering excellence as its competitive edge confronted its own turning point. Technology investment had been steady but narrowly focused on efficiency. As competitors began integrating digital tools into product design, leadership realized that its definition of technology needed to expand. The company launched a modernization initiative not to reduce cost but to enable faster innovation. Engineers were trained in data analytics. Product teams began collaborating with IT and marketing. The effort started as a technology project but quickly evolved into a cultural and strategic shift. Technology became the lens through which the organization learned to move faster and work together, setting the foundation for what would later become a distinct advantage.

In all three cases, technology provided the initial spark, but real progress came only when it connected to leadership and strategy. Technology can modernize systems and accelerate decisions, but it cannot define ambition or coherence. Without leadership to establish intent and strategy to focus effort, technology becomes a collection of tools rather than an enabler of advantage.

Technology-led beginnings are powerful because they are visible. They demonstrate that change is possible and that momentum can be built from capability outward. Yet they also remind leaders of the limits of technology in isolation. Every organization, whether starting, renewing, or continuing the pursuit of modern advantage, must evolve through this approach. Distinction emerges only when technology moves in rhythm with leadership and strategy, creating clarity, coherence, and pace that the market can feel.

In some cases, urgency compresses sequence altogether, forcing leadership, strategy, and technology to evolve at once.

Modernizing All at Once

From time to time, organizations find themselves modernizing leadership, strategy, and technology all at once. The decision is rarely deliberate. It emerges when pressure builds and the cost of waiting becomes greater than the cost of change. These efforts are not demonstrations of boldness; they are responses to urgency. Modernizing all at once is what happens when the environment demands motion faster than sequence can allow.

This path can create early momentum, but it also tests discipline. When all three variables evolve simultaneously, clarity is difficult to sustain. Leadership may move faster than technology can support. Strategy may advance before new capabilities are ready to deliver. Each discipline begins to change, but not always in rhythm. The challenge is not how to move but how to keep direction visible while everything moves at once.

A telecommunications company faced this reality when new competitors began eroding market share almost overnight. Leadership understood that the existing strategy and infrastructure could not respond quickly enough. Waiting to modernize one discipline at a time would have meant losing relevance entirely. The company launched concurrent efforts to better align leadership competencies, redefine market priorities, and overhaul its technology platform. The first months were chaotic. Progress in one area exposed gaps in another. Yet through consistent communication and measured pacing, coherence began to form. The organization did not slow the environment; it learned to move with it.

A global humanitarian organization experienced a different kind of pressure. A series of major crises exposed structural and technological limitations that slowed its response. Legacy systems, decentralized authority, and uneven decision-making capacity undermined speed and trust. The organization moved to modernize leadership accountability, operational strategy, and technology infrastructure simultaneously. The work was messy, and fatigue was high, but the urgency of the mission kept focus intact. Within a year, coordination improved dramatically, and field operations became faster and more adaptive. The organization learned that even in chaos, coherence depends on connection, not control. Response times shortened significantly, and donor confidence returned, showing that coherence strengthens even under extreme pressure.

A public-sector agency responsible for citizen services faced its own version of this challenge. Incremental improvement had failed to meet rising expectations, and public confidence was slipping. With pressure intensifying, leaders decided to act on all three fronts at once. They clarified accountability, redefined strategic outcomes, and accelerated a digital modernization program that had stalled for years. The effort was uneven and fatigue was real, but urgency created focus. By managing tempo instead of chasing perfection, the agency regained both momentum and credibility. Citizen satisfaction scores improved, and employee engagement rose, validating that speed aligned with purpose restores trust.

Modernizing all at once is rarely comfortable, but it can be effective when guided by the principles of the modern approach. The organizations that succeed under pressure do not move faster by chance. They impose rhythm on complexity. They maintain clarity, sequence decisions deliberately, and stay connected even when the pace feels unmanageable. Through that discipline, the approach begins to take hold, and the organization starts to build what will become its distinct advantage.

Addressing all three variables at once does not remove the need to synchronize them later. It only accelerates their movement toward that point. Modernization builds capability; synchronization converts it into coherence. Even when leadership, strategy, and technology evolve together under pressure, advantage lasts only when the organization finds rhythm. Coherence is not achieved by intensity; it is sustained through connection, discipline, and the willingness to keep adapting at speed.

Across these different beginnings, the outcome is consistent. Progress becomes durable only when motion turns into rhythm, and rhythm, practiced consistently within context, becomes the foundation of a distinct advantage.

The Pattern Beneath the Pursuit

Across these examples, one pattern stands out. Every organization begins or renews the pursuit of modern advantage from its own reality, yet progress always follows the same logic. The variable that moves first may differ, but the forces that shape motion are constant.

Clarity creates confidence. Connection creates coherence. Coherence creates rhythm. Once that rhythm appears, the pursuit sustains itself.

This is what it means to start where you are. It is not a statement of limitation; it is a recognition of truth. No organization begins the pursuit under perfect conditions. Each works from the pressures, priorities, and opportunities it already faces. Some start from strength, others from strain. What matters is not the point of origin, but the willingness to move and to connect the disciplines that create motion.

The modern approach exists to make that motion possible. It provides a way to translate difference into direction, speed into rhythm, and effort into progress. It allows organizations to act with urgency without losing coherence. Leadership establishes intent, strategy shapes focus, and technology accelerates both. When these disciplines move together in rhythm, the organization begins to build what becomes its distinct advantage, the visible expression of coherence practiced in context.

Starting where you are does not describe a phase; it describes a posture. It is how organizations remain relevant in a world that does not pause. The pursuit of advantage never truly starts or stops; it only speeds up, slows down, or changes form. The organizations that sustain a distinct advantage are those that keep beginning again, connecting what they know with what they must still learn.

Chapter Summary

Every organization begins or renews the pursuit of modern advantage from a different place. Some start from disruption, others from opportunity, and some from the realization that progress has slowed. These differences matter, but they do not alter the fundamentals of pursuit. The modern approach works from wherever an organization happens to be because it is built for motion. It transforms difference into direction, complexity into rhythm, and activity into coherence.

When practiced within a unique environment, that approach becomes a distinct advantage.

The examples in this chapter show that the pursuit can begin in many ways. Some organizations start with leadership, re-establishing clarity and pace. Others begin with strategy, redefining focus and choice. Many now begin with technology, building capability that reconnects ambition with execution. A few, under pressure, attempt to evolve all three disciplines at once. Their starting points differ, but their outcomes converge. Progress only becomes sustainable when leadership, strategy, and technology move together in rhythm, turning motion into momentum and momentum into distinction.

Starting where you are is not a matter of choice; it is a condition of the modern environment. The pursuit of advantage no longer waits for alignment or perfection. It begins wherever readiness and urgency meet, then grows stronger through connection. The next chapter, *Chapter 14: Owning the Pursuit,* turns from proof to responsibility. It explores how organizations sustain momentum once the pursuit is underway and how leaders ensure that coherence becomes a lasting habit rather than a temporary achievement.

CHAPTER 14:
OWNING THE PURSUIT

Where Ownership Begins

Every pursuit needs an owner. The modern approach to advantage does not build or sustain itself, and coherence does not appear by chance. They require stewardship. The responsibility to lead the pursuit belongs first to those who shape direction, set pace, and define standards. Ownership begins with leadership.

For most organizations, ownership starts with recognition. It begins when leaders acknowledge the need to pursue advantage deliberately within their own environment. Recognition is the moment when intent becomes visible. It is the understanding that the pursuit is both relevant and attainable, that advantage is not reserved for others or for better conditions, but is possible here and now. Leaders who recognize this truth shift from reacting to events to shaping them. They see pursuit not as an optional initiative but as the defining responsibility of leadership.

Owning the pursuit also requires prescription. Leaders must define the approach their organization will use to create coherence and sustain rhythm. The most effective prescription is one that connects leadership, strategy, and technology in motion, turning awareness into progress. Prescription is not about control; it is about clarity. It defines what progress means in context and creates shared understanding of how those disciplines should move together.

Without that shared definition, motion fragments into effort rather than advantage.

Resilience gives ownership its character. No pursuit moves in a straight line, and few conditions stay stable long enough to make progress easy. Leaders who own the pursuit accept disruption as part of the process. They build capacity to adjust without losing direction. Resilience does not mean ignoring fatigue; it means responding to it. It is the steady conviction that renewal is possible even when pressure rises.

Consider an energy company navigating the transition to renewables. The shift required new technology, new partnerships, and a new operating model. Early progress faltered when competing priorities fractured focus. The CEO gathered her team and reframed ownership around rhythm, not projects. She reminded them that progress was not measured by the number of initiatives launched but by the consistency with which leadership, strategy, and technology moved together. Within months, decisions became faster, and collaboration strengthened. The company rediscovered its momentum not by changing the goal but by renewing ownership of the approach.

Ownership is not an act of possession. It is a posture of accountability that extends across time. The leaders who build and sustain a distinct advantage understand that their role is to hold the pursuit steady while the environment shifts. They start wherever clarity allows and keep moving when conditions make progress difficult. The measure of leadership is not how often they begin but how reliably they continue.

Ownership begins with recognition, prescription, and resilience. It is the decision to see clearly, define deliberately, and endure confidently. Every organization must choose that posture for itself—because the pursuit of advantage cannot be delegated. When leaders hold the approach with that discipline, the result is coherence that endures and a distinct advantage that lasts.

Making Coherence Endure

Ownership means little if it cannot be sustained. Once leaders define the approach that connects leadership, strategy, and technology in motion, the next task is to make that connection permanent. Coherence must outlast a moment of clarity or a burst of urgency. It must become part of how the organization works every day.

Once ownership is established, sustaining it becomes the challenge. The modern approach depends on rhythm, not events. Many organizations still treat transformation as a cycle of large initiatives separated by periods of maintenance. That pattern no longer fits the pace of the modern environment. The work of coherence cannot pause while conditions change. It must continue through smaller, more frequent renewals that keep the organization aligned and relevant. Sustaining advantage is less about the scale of change and more about the consistency of connection.

Establishing this approach once creates the framework for continuity. The disciplines of leadership, strategy, and technology provide the structure. The rhythm between them provides the stability. When each discipline reinforces the others, coherence stops depending on individual leaders or single programs. It becomes self-correcting. Over time, decisions stay aligned even as personnel and priorities evolve.

Consider a global logistics company that redefined its operating model around continuous renewal. In the past, leadership had launched major transformation efforts every few years, each one beginning with excitement and ending with exhaustion. The company replaced those cycles with a lighter, continuous model. Every quarter, leaders reviewed alignment across the three disciplines, clarity of direction, strategic focus, and technological capability. Adjustments were small but regular. Within a year, performance improved, and employee engagement rose. The organization no longer relied on momentum from projects; it drew stability from rhythm.

The benefit of this endurance is resilience that compounds. When coherence becomes a constant, adaptation becomes easier. The organization learns faster, reacts sooner, and invests with greater confidence. Opportunities are recognized earlier, and disruptions lose their power to disorient. The pursuit of advantage stops being a response to change and becomes a force that shapes it. When practiced with this consistency, the approach produces the conditions that make a distinct advantage possible.

Making coherence endure also requires discipline in governance. The pursuit of advantage must be visible in how decisions are made and how progress is measured. Leaders who own coherence treat it as a permanent agenda item, not a temporary priority. They align metrics, incentives, and accountability to reinforce this approach. When coherence is embedded in the language of performance, it stops being an aspiration and becomes a practice.

Sustaining a distinct advantage is not about repeating the same actions; it is about preserving the conditions that allow rhythm to continue. The most effective organizations view coherence as an operating system, not a strategy. It is the framework through which new strategies, technologies, and leadership styles can emerge without fragmenting progress.

Making coherence endure means establishing this approach once and returning to it constantly. The reward is not only stability but acceleration. Coherence multiplies the value of every investment, decision, and initiative. It turns renewal into routine and progress into performance. When coherence becomes habit, advantage becomes renewable, and the organization's distinct advantage becomes self-sustaining.

The Discipline to Sustain

Every organization that commits to the pursuit of modern advantage faces the same reality: the work is demanding. This approach asks leaders first to build excellence within each discipline and then to bring those disciplines into rhythm. That sequence takes time. Leadership must strengthen its credibility and focus. Strategy must clarify direction and coherence. Technology must build capability that can sustain pace. Only when those foundations exist can true alignment begin. The early stages are where most organizations struggle not because the concept is flawed, but because the discipline it demands must be learned before it can become instinctive.

Sustainability is the difference between intention and achievement. The pursuit of advantage will compete with every other priority for attention, resources, and patience. The temptation to defer or dilute the work is strong, especially when results seem distant. Leaders who sustain the pursuit treat coherence as the priority that shapes all others. They understand that every initiative, investment, and policy is more effective when the organization moves as one. Coherence is not an additional objective; it is the condition that makes every objective possible.

The early effort is the heavy lifting. Building excellence within and connection across requires new habits of leadership, new forms of coordination, and often a redefinition of how decisions are made. The work can feel uncomfortable because it challenges systems and behaviors that once produced success. But it is this very discomfort that signals progress. The pursuit begins to take hold when leaders and teams embrace that discomfort as evidence that the organization is growing stronger.

Remaining diligent does not mean working alone. Many organizations benefit from outside perspective during this stage not because they lack competence, but because they recognize the value of objectivity. External partners can help reveal where depth is thin or where connections are weak. Their perspective can accelerate progress and reduce the risk of fatigue. Seeking assistance is not an admission of deficiency; it is an act of commitment. It signals that leadership intends to see the approach through to the point where connection becomes self-sustaining.

Consider a financial services firm that set out to embed the modern approach after a period of inconsistent performance. The leadership team knew that technology investments had outpaced strategy, and that strategy had outpaced communication. Progress was slow until the organization brought in a partner to help realign its priorities. The engagement focused on strengthening the fundamentals of each discipline before re-establishing the rhythm between them. Within a year, coordination improved, duplication declined, and results became more predictable. The firm did not outsource the pursuit of advantage; it strengthened its ability to lead it.

Sustaining advantage is not about speed; it is about conviction. The work of building excellence within and connection across will never be effortless, but the organizations that persevere through that effort reach a point where discipline becomes strength and rhythm replaces strain. They earn the right to operate with confidence because they have built the structure that allows renewal to happen naturally.

The pursuit of advantage must be viewed as the priority overarching all other priorities. It is the framework that amplifies performance, accelerates learning, and protects relevance. Treating it as anything less eventually costs more than the effort required to embed it. The path is demanding, but what pursuit could possibly matter more? Without excellence, connection has nothing to unify. Without connection, advantage cannot last. The organizations that thrive are the ones that choose to do the work, knowing that the effort required to build a distinct advantage is always less than the cost of losing it.

The Decision to Begin

Every organization that aspires to modern advantage faces a choice: to begin or to wait. Waiting often feels justified. There is always another initiative to complete, another crisis to manage, another quarter to deliver. The pursuit of advantage is easy to postpone because its urgency is rarely loud. Yet postponing the work does not make it easier. It only allows complexity to grow and opportunities to pass. The hardest part of the pursuit is not the effort it requires but the decision to begin.

Most leaders understand the value of advantage. They believe in focus, adaptability, and distinction. What holds them back is not disbelief but doubt. The modern environment moves quickly, and the work of building excellence within and connection across can feel daunting. The speed of change can make the pursuit seem impossible to start. But speed is exactly why it must start. In a world that will not slow down, rhythm becomes the only form of control that remains.

Beginning is the moment when leadership turns conviction into motion. It means deciding that progress matters more than perfection. It means accepting that the work will be demanding and that discomfort is proof of movement. Leaders who begin the pursuit do not wait for conditions to align; they create alignment through action. They learn as they go, building credibility within and connection across until rhythm emerges. The pursuit rewards those who act first and refine later.

Every organization that has succeeded in embedding the modern approach has faced the same hesitation. Each one began from a place of uncertainty and moved forward anyway. They started with small steps, not master plans. They focused on one area of strength and pulled the others toward it. Over time, the work became lighter, the rhythm steadier, and the results clearer. The difference between those who sustained advantage and those who did not was never knowledge; it was initiation.

Consider a healthcare network that had delayed its pursuit of advantage for years, waiting for clearer market conditions and a more stable budget. By the time leadership finally began, the market had moved again, and the organization's delay had become its greatest obstacle. The turning point came when the CEO reframed the pursuit as a responsibility, not an option. She told her team that the effort might take years, but the cost of inaction would last longer. That statement changed the tone of leadership. The organization began with what it had—committed people, partial clarity, and the willingness to act. Within two years, progress was visible. Renewal had replaced hesitation, and the organization had begun to build what would become its distinct advantage.

The decision to begin is not an admission of unpreparedness; it is an expression of intent. It is the recognition that pursuit is a leadership choice, not a situational advantage. The modern approach is not a single leap but a sequence of deliberate steps. Each one builds credibility, connection, and capability until excellence becomes habit. Leaders who start discover that what once seemed heavy becomes manageable through motion, and that motion itself becomes momentum.

The pursuit of advantage will always demand focus, energy, and belief. But nothing valuable in leadership has ever required less. The greater risk lies in waiting. Every day spent debating the start is a day spent strengthening inertia. The work is hard, but what could possibly be more important? Advantage is not achieved by knowing what to do but by deciding to do it. When that decision is made and sustained through the modern approach, it becomes the first act in creating a distinct advantage.

Chapter Summary

The pursuit of modern advantage begins with ownership. Leaders must recognize that the approach will not build or sustain itself, and that their first responsibility is to lead the work deliberately. Ownership is a posture of accountability that combines recognition, prescription, and resilience. It requires clarity of intent, definition of method, and the conviction to act even when conditions are uncertain.

Sustained advantage depends on discipline. The modern approach turns strength within and connection across into a living rhythm that keeps organizations aligned, adaptable, and credible. When that rhythm becomes habit, progress compounds and renewal becomes instinctive. The effort required to embed the pursuit is significant, but the reward is lasting stability and greater freedom to lead.

The work is demanding, but it is also defining. The pursuit of advantage is not a program to complete but a standard to uphold. It is the framework that amplifies performance, accelerates learning, and protects relevance. Every leader and every organization must ultimately choose to begin and to keep beginning. Those who do the work, who build excellence within and synchronize strength across, earn more than performance. They earn coherence that endures, and through it, a distinct advantage that lasts.

CONCLUSION

When the World Changes, So Must the Pursuit of Advantage

Every book ends, but the work this one describes does not. If there is a single truth that endures through every chapter, it is that the world has changed, and with it, the pursuit of advantage. That change did not come from theory or preference; it came from pressure.

For much of the last century, organizations operated in a world that rewarded control. Advantage could be built through size, defended through efficiency, and displayed through brand and distribution. The conditions of competition were predictable enough that repetition created strength. If you were disciplined, consistent, and patient, you could build advantage that lasted for decades.

Those same instincts now create drag. The environment that once rewarded steadiness now demands motion. Markets shift faster than plans can be refreshed. Customers move across categories that used to be separate. Technology dissolves boundaries that once defined industries. Information travels instantly, and expectation travels with it. Scale, brand, and distribution still matter, but they are outcomes, not engines.

The world has become too complex, too fast, and too connected for the old pursuit to hold. Size, reputation, and efficiency can still measure success, but they can no longer create it. They are evidence of strength, not its source. Organizations that still chase them as causes find themselves competing on yesterday's terms, optimizing what used to work instead of building what will.

That is why a modern approach to advantage exists. It is the only approach that matches the world as it now operates. The modern approach replaces the illusion of control with the discipline of coherence. It trades the pursuit of scale for the pursuit of strength that compounds over time. Advantage now depends on two forms of mastery: excellence within the disciplines that create value and synchronization across them to sustain momentum.

This approach is not simple. It requires leaders to think in motion instead of in parts, to trade certainty for clarity, and to make progress a continuous condition rather than an occasional event. The pace of change is permanent; complexity is ordinary. The only way to keep up is to build and preserve both depth and rhythm, to sustain excellence within while keeping those strengths connected in motion. The modern approach makes that possible.

Leadership, strategy, and technology define this approach, but before they can work together, each must evolve on its own. Leadership can no longer rely on hierarchy or authority to create alignment; it must generate clarity in motion. Strategy can no longer depend on prediction or static plans; it must become an instrument of learning, translating direction into adaptable choices. Technology can no longer sit apart as infrastructure; it has become the environment in which the business operates, shaping how decisions are made and performance is sustained.

Only once these disciplines evolve individually can they interact effectively. The modern pursuit of advantage depends on their balance, excellence inside and rhythm across. When leadership, strategy, and technology move together, they create coherence that keeps an organization steady while the world accelerates around it.

That balance does not promise ease. It demands practice, patience, and humility. It requires leaders to modernize the true causes of performance rather than admire the old symptoms of success. It insists that organizations replace the pursuit of control with the pursuit of connection. Those who commit to this work discover that the effort is not only worth it but essential. The modern approach is not another method; it is the foundation of how advantage now behaves.

Without it, organizations may still perform, but they will not progress. They will move faster for shorter bursts and work harder for smaller gains. They will confuse activity with renewal and motion with meaning. They will appear healthy until the pace of their environment exposes how fragile they have become. In contrast, organizations that embrace the modern approach find a steadier form of progress. They learn to convert speed into rhythm and complexity into coherence. Their advantage is not that they face fewer challenges, but that they can move through them.

This book was written to make that shift visible. *The modern approach to advantage is not an alternative to what came before; it is the only approach that fits what comes next.* The world has replaced the luxury of control with the necessity of connection. It rewards those who evolve faster than their surroundings and punishes those who cannot. The question for every organization is not whether the work is difficult, but whether it is harder to change or harder to stand still.

The pages that follow in this conclusion are not about technique. They are about conviction: what happens when the modern approach stops being something to learn and starts becoming something to live. While the work is demanding, it is also the only work that leads to advantage that endures. In a world that moves this quickly, ordinary will always be temporary.

The Modern Pursuit

Understanding the modern approach to advantage is the easy part. Living it is where the work begins. The organizations that succeed are not the ones that memorize a process, but those that build excellence within leadership, strategy, and technology, then connect those strengths in motion. They begin wherever they have the most leverage and let progress in one discipline pull the others forward.

The modern pursuit is not a theory to apply; it is a discipline to build. It depends on two connected forms of mastery. The first is excellence within each of the three core disciplines. Leadership, strategy, and technology must each achieve depth that earns confidence inside the organization. The second is synchronization across them so the excellence within reinforces rather than competes. Depth gives the organization credibility; rhythm turns that credibility into momentum. Both are essential.

At first, this work feels mechanical. Leaders modernize their own domains. Leadership evolves from authority to clarity, building adaptability and trust. Strategy shifts from prediction to learning, defining focus while leaving room to adjust. Technology moves from a set of tools to the connective environment of the enterprise. Each change deepens the foundation of advantage. Excellence within the disciplines creates the stability that rhythm later amplifies.

Once that depth is established, coherence begins to emerge. Teams notice that progress depends less on intensity and more on consistency. Meetings shorten because decisions are made with shared context. Priorities align not through oversight but because excellence within the disciplines gives rhythm something solid to connect. Progress starts to feel steady rather than forced.

This is the balance at the heart of the modern pursuit. Excellence within gives the organization its strength; synchronization across gives it movement. Without excellence, rhythm collapses under pressure. Without rhythm, excellence loses direction. The pursuit of advantage requires both, practiced deliberately until they become inseparable.

As organizations strengthen this balance, the work of leadership changes. It becomes less about instruction and more about interpretation. Leaders sense how well the disciplines are working together, noticing when one starts to outpace the others, when strategy has moved ahead of capability, when technology advances faster than adoption, or when communication loses clarity. Their task is not to control the pace but to preserve equilibrium, depth inside and rhythm across.

The same evolution happens within strategy. Modern strategy is no longer a plan to defend but a living design that keeps coherence intact. It defines where to focus but stays flexible about how. It brings discipline to direction and creates alignment that survives change. Strategy done well reflects both mastery and motion, the ability to hold focus without freezing in place.

Technology converts both elements, excellence and rhythm, into delivery. It carries leadership's direction and strategy's priorities across the organization at the speed markets now require. It provides visibility that reveals whether rhythm is being sustained and whether excellence within each discipline remains current. Technology becomes the translator of coherence, turning depth into motion and making alignment visible through execution.

When excellence within and rhythm across reach equilibrium, progress becomes renewable. The organization no longer waits for transformation cycles to generate energy. Renewal becomes a natural outcome of daily work. People know what to expect from leadership, understand how strategy connects to purpose, and trust that technology will enable rather than distract. Clarity and connection feed each other.

It takes time for this balance to feel natural, but once it does, it changes how progress feels. Work becomes deliberate instead of reactive. Teams move with confidence because they understand how their part strengthens the whole. Meetings are shorter, decisions faster, and results steadier. The organization learns to adapt without losing coherence and to modernize without starting over.

This is what defines the modern pursuit: the constant renewal of excellence and rhythm, credibility and connection. The pursuit of advantage no longer rewards those who are simply fast or large; it rewards those who are both capable and coherent, both skilled and synchronized. When excellence within and rhythm across reinforce one another, progress becomes a habit and advantage becomes a state of motion that endures.

From Understanding to Practice

The modern approach to advantage is universal in principle but personal in practice. Every organization faces the same forces of pace, transparency, and complexity, yet each must find its own way to move through them. The approach does not prescribe; it enables. It gives shape to motion without dictating form. Advantage becomes distinct only when the same modern approach is practiced within the realities an organization already owns.

The work begins with excellence. The first task of modernization is depth within leadership, strategy, and technology. Each discipline must reach a level of credibility strong enough to stand on its own before it can reinforce the others. Leadership earns trust through clarity and consistency. Strategy proves focus through disciplined choice. Technology delivers reliability through capability that endures under pressure. This excellence within creates the foundation for coherence. Without that depth, rhythm has nothing to connect.

Once excellence is present, rhythm turns it into progress. Connection across disciplines does not replace mastery; it multiplies it. When leadership, strategy, and technology operate in balance, the organization begins to move with coherence that others can sense but cannot easily copy. That motion, shaped by context and powered by credibility, is what makes advantage distinct.

The path looks similar across organizations, though it never feels the same. A financial institution may find rhythm through focus, letting clarity of strategy pull leadership and technology into motion. A manufacturing company may begin with technology, using integration to restore coherence between decision-making and delivery. A services firm may start with leadership, aligning culture and communication before turning outward. What matters is not where the pursuit starts but how excellence within and rhythm across develop together. Progress emerges when both strengthen in concert.

As the modern approach takes hold, it stops feeling deliberate and starts becoming natural. Leaders stop describing modernization as a program; they simply practice it. Teams no longer wait for clarity from above; they create it through collaboration. Technology fades into the background, carrying information and intent through the organization as part of how work happens. The pursuit of advantage no longer depends on initiatives or transformations. It continues through daily habits that keep depth current and rhythm intact.

When the modern approach takes root, distinction appears without being designed. The difference between one organization and another lies not in their models but in how they express the same approach. Every company carries its own mix of history, culture, and constraint. These factors shape how coherence feels and how excellence shows up. In one enterprise, distinct advantage might reveal itself as speed. In another, as reliability. In another, as trust. The method is universal, but its expression is personal.

This is the strength of the modern approach. It creates consistency without conformity. It allows an organization to stay grounded in shared principles while adapting them to local realities. It makes alignment portable and renewal repeatable. Excellence within keeps the organization credible; rhythm across keeps it alive. Together, they form a pattern of progress that no competitor can replicate because that pattern reflects how one organization has learned to think, decide, and move together.

Over time, this balance becomes part of identity. The practices that once felt intentional become instinctive. Excellence within continues to evolve as rhythm across matures. Renewal stops being a project and becomes a reflex. At that point, the approach has done its work. It has turned coherence into character and rhythm into reputation.

The lesson is simple but demanding. A distinct advantage is not achieved by building something different; it is achieved by living something universal until it becomes personal. The modern approach gives every organization the same foundation: excellence within leadership, strategy, and technology, strengthened by rhythm across them. When that balance is practiced within a specific environment, its markets, its culture, and its choices, it becomes distinct by definition. Advantage remains modern in method but unmistakably its own in form.

Advantage Is Human Work

Every pursuit eventually comes down to people. The modern approach to advantage may rely on leadership, strategy, and technology, but none of these forces act on their own. Coherence is built through human judgment. Renewal depends on human attention. Advantage, at its core, is human work.

The most advanced ideas or tools matter only to the extent that people use them well. Excellence within begins as personal discipline, leaders who practice clarity, consistency, and adaptability until they become instinct. Strategy achieves coherence when people connect purpose to action, making focus visible through daily choices. Technology supports progress when people treat it as an extension of judgment rather than a replacement for it. The approach lives through behavior before it is visible in results.

This is what separates organizations that endure from those that fade. The strongest do not rely on a single moment of alignment; they sustain it through conduct. Their leaders create trust by staying present through change. Their teams listen for rhythm in how work feels, how decisions are made, how clarity is shared, and how progress repeats. Renewal becomes everyone's responsibility.

That rhythm remains fragile because it is human. It can drift under pressure, noise, or fatigue. The work of leadership is to protect it. Leaders maintain coherence by restoring focus, communicating transparently, and showing steadiness when others hesitate. Their task is not control but calibration, keeping the organization balanced between depth and motion.

Leadership in this environment demands humility and precision. It begins with awareness, knowing how the organization feels, where timing is slipping, and where clarity has thinned. It requires curiosity to keep learning as conditions evolve, and patience to let rhythm take hold before declaring results. The ability to remain coherent under stress has become the defining mark of modern leadership.

Advantage is never built from the outside in. Markets, investors, and technologies set the pace, but advantage is created in the choices people make together each day. Coherence cannot be delegated. It must be led, modeled, and reinforced. Excellence within is sustained by individual discipline; rhythm across is sustained by collective trust. Both rely on the same truth, advantage endures only when people choose to move together.

When that choice becomes habit, progress feels natural. Teams begin to anticipate one another's decisions. Leaders find that alignment continues even when they are not directing it. Work feels lighter because rhythm has replaced urgency. What began as a method becomes culture. What began as intention becomes identity.

Every generation believes its conditions are more complex than the last, and perhaps they are. Yet the responsibility remains the same. Advantage is not maintained through control but through connection. It is built by people who bring order to motion and purpose to change. The modern approach gives them the means to do so. It allows clarity to move at the pace of the world and coherence to hold through every cycle of disruption.

Advantage will always depend on technology and strategy, but it will always begin and end with people. Systems can record rhythm, but only people can create it. In every environment, under every condition, advantage remains human work, the steady practice of keeping purpose clear, motion coherent, and progress alive.

What Endures

Every era believes its version of progress is final. Each imagines that advantage can be secured through effort alone. Time always proves otherwise. Markets move, technology resets the pace, and what once felt stable begins to erode. Yet through every shift, the same truth persists: coherence outlasts control, and renewal outlasts size.

The organizations that endure understand this difference. *They treat advantage as a living pursuit, not a finished achievement.* They invest in depth before speed and rhythm before reach. They strengthen leadership, strategy, and technology not as separate disciplines but as connected sources of clarity. They know that excellence within creates confidence, and that rhythm across turns that confidence into progress. They repeat this work until it becomes the normal condition of how they operate.

For them, advantage is not a race to win but a rhythm to sustain. It grows through learning, patience, and the quiet repetition of practices that keep coherence intact while everything else changes. The world will continue to accelerate, and competition will remain relentless. Yet in motion, steadiness becomes rare, and in complexity, coherence becomes visible. That visibility is what creates trust.

This is what endures. The modern approach to pursuing competitive advantage is not a reaction to disruption; it is the discipline of moving through it. It is how organizations hold focus while the ground shifts and how leaders translate intention into rhythm that others can feel. The pursuit is demanding, but it defines what leadership now means. It is the difference between those who manage the moment and those who create the future.

Every advantage is temporary for those who treat it as possession. It becomes enduring only for those who treat it as practice. The work described in these pages never ends, but it rewards those who continue it. The world will always move faster than comfort allows, yet within that motion, coherence and depth remain possible. Those who preserve both will not only adapt; they will stand apart.

ABOUT THE AUTHOR

For more than three decades, Tom Mawhinney has helped organizations compete, grow, and stay relevant in environments defined by speed, complexity, and constant change. His work as a board member, senior executive, consultant, and entrepreneur has spanned industries and economic cycles—giving him a front-row view of how leadership, strategy, and technology have evolved into the defining forces of modern competitive advantage.

A recognized expert in contemporary leadership, strategic coherence, and the role of technology in shaping performance, Tom has advised executives, founders, and boards around the world. His work bridges timeless principles with the realities of emerging technologies and shifting stakeholder expectations—helping leaders build clarity, coherence, and traction when conditions refuse to stand still.

Guided by deep curiosity and a commitment to continuous learning, Tom remains at the forefront of how organizations create and renew advantage. Through his writing, speaking, and advisory partnerships, he equips leaders to modernize their capabilities, synchronize the forces that drive performance, and pursue a distinct advantage in an ever-accelerating world.

Email: Tom@TomMawhinney.io

Website: www.TomMawhinney.io

LinkedIn: https://www.linkedin.com/in/tommawhinney

DID YOU ENJOY THIS BOOK?

If you enjoyed reading this book, you can help by suggesting it to someone else you think might like it, and **please leave a positive review** wherever you purchased it. This does a lot in helping others find the book. We thank you in advance for taking a few moments to do this.

THANK YOU

MORE TITLES FROM THE AUTHOR

AUTHOR

The Contemporary Leader: The Modern Skills Required to Lead, Adapt, and Succeed in Today's Marketplace

The Contemporary Board Member: The 15 Essential Leadership Skills Reimagined for the Boardroom

The Contemporary Executive: The 15 Essential Leadership Skills Activated in the Executive Suite

The Contemporary Entrepreneur: The 15 Essential Leadership Skills Applied to the Entrepreneurial Journey

CONTRIBUTING AUTHOR

The AI Revolution: Thriving Within Civilization's Next Big Disruption

www.ingramcontent.com/pod-product-compliance
Lightning Source LLC
Chambersburg PA
CBHW071602210326
41597CB00019B/3359